This series aims to report new developments in physical research and teaching – quickly, informally, and at a high level. The type of material considered for publication includes:

1. Preliminary drafts of original papers and monographs

2. Lectures on a new field, or presenting a new angle on a classical field

3. collections of seminar papers

4. Reports of meetings

Texts which are out of print but still in demand may also be considered if they fall within these categories.

The timeliness of a manuscript is more important than its form, which may be unfinished or tentative. Thus, in some instances, proofs may be merely outlined and results presented which have been or will later be published elsewhere.

Publication of *Lecture Notes* is intended as a service to the international physical community, in that a commercial publisher, Springer-Verlag, can offer a wider distribution to documents which would otherwise have a restricted readership. Once published and copyrighted, they can be documented in the scientific libraries.

Manuscripts

Manuscripts are reproduced by a photographic process; they must therefore be typed with extreme care. Symbols not on the typewriter should be inserted by hand in indelible black ink. Corrections to the typescript should be made by sticking the amended text over the old one, or by obliterating errors with white correcting fluid. The figures (in the original size) ready for reproduction should be inserted into the text. Should the text, or any part of it, have to be retyped, the author will be reimbursed upon publication of the volume. Authors receive 50 free copies.

The typescript is reduced slightly in size during reproduction, therefore a large size of type should be used; best results will not be obtained unless the text on any one page is kept within the overall limit of 18 x 26.5 cm (7 x 10½ inches). **The publishers will be pleased to supply on request special stationery with the typing area outlined.**

Manuscripts in English, German or French should be sent to Springer-Verlag, 6900 Heidelberg, Postfach 1780.

Die „*Lecture Notes*" sollen rasch und informell, aber auf hohem Niveau, über neue Entwicklungen in der Physik berichten. Zur Veröffentlichung kommen:

1. Vorläufige Fassungen von Originalarbeiten und Monographien.

2. Spezielle Vorlesungen über ein neues Gebiet oder ein klassisches Gebiet in neuer Betrachtungsweise.

3. Seminarausarbeitungen.

4. Vorträge von Tagungen.

Ferner kommen auch ältere vergriffene spezielle Vorlesungen, Seminare und Berichte in Frage, wenn nach ihnen eine anhaltende Nachfrage besteht.

Die Beiträge dürfen im Interesse einer größeren Aktualität durchaus den Charakter des Unfertigen und Vorläufigen haben. Sie brauchen Beweise unter Umständen nur zu skizzieren und dürfen auch Ergebnisse enthalten, die in ähnlicher Form schon erschienen sind oder später erscheinen sollen.

Die Herausgabe der „*Lecture Notes*" Serie durch den Springer-Verlag stellt eine Dienstleistung an die physikalischen Institute dar, indem der Springer-Verlag für ausreichende Lagerhaltung sorgt und einen großen internationalen Kreis von Interessenten erfassen kann. Durch Anzeigen in Fachzeitschriften, Aufnahme in Kataloge und durch Anmeldung zum Copyright sowie durch die Versendung von Besprechungsexemplaren wird eine lückenlose Dokumentation in den wissenschaftlichen Bibliotheken ermöglicht.

Lecture Notes in Physics

Edited by J. Ehlers, Austin, K. Hepp, Zürich and
H. A. Weidenmüller, Heidelberg
Managing Editor: W. Beiglböck, Heidelberg

7

R. Balescu, J. L. Lebowitz,
I. Prigogine, P. Résibois, Z. W. Salsburg

Lectures in Statistical Physics

From the Advanced School for Statistical
Mechanics and Thermodynamics
Austin, Texas USA

Compiled by W. C. Schieve, M. G. Velarde, A. P. Grecos
Center for Statistical Mechanics and Thermodynamics,
University of Texas, Austin, Texas USA

Springer-Verlag
Berlin Heidelberg GmbH 1971

ISBN 978-3-540-05418-4 ISBN 978-3-540-36535-8 (eBook)
DOI 10.1007/978-3-540-36535-8

Originally published by Springer-Verlag Berlin Heidelberg New York in 1971.

Library of Congress Catalog Card Number 78-155594.

PREFACE

These lectures are taken from the Advanced School for Statistical Mechanics and Thermodynamics organized by the Center for Statistical Mechanics of the University of Texas at Austin,(Professor Ilya Prigogine director). All lectures, except for those by Professor Lebowitz, are from the first school held in Spring 1969. Those by Professor Lebowitz are from the following year.

We feel a deep sadness at the death of Professor Zevi Salsburg during the preparation of this manuscript. His lectures were assembled from tape recordings and lecture notes. We hope these notes reflect in a small way his quality as a great teacher which we are sure his students and colleagues at the first Advanced School sincerely appreciated.

The lectures in this series form a natural sequence. Professor Prigogine first discusses the new developments in the macroscopic theory of non-equilibrium thermodynamics in the non-linear domain, particularly the dissipative structures", order appearing in far from equilibrium states. The theme of order continues in Professor Salzburg's review of equilibrium properties of phase transitions with particular emphasis on the question of long range order in one, two and three dimensional space. Dynamical effects are then reviewed by Professor Resibois in his lecturers on more recent developments. Here he discusses, for example, dynamical scaling, the semi-microscopic theory of Kadanoff and Swift. He then discusses a microscopic model reviewing his work on Heisenberg spin systems using the full techniques of non-equilibrium statistical mechanics. Professor Lebowitz turns to deep and fundamental questions when he finishes the theme of order by concisely reviewing the rigorous statistical mechanical proofs of the existence of the thermodynamic limit in equilibrium, and also the recent work on the existence of solutions to equations of motion in the limit of an infinite number of particles. The final article by Professor Balescu is naturally related to those of Resibois and Lebowitz being a review of the formulation of non-equilibrium statistical mechanics from the Liouville equation with particular emphasis on the projection of asymptotic general kinetic equations.

Finally, we would like to thank the National Science Foundation and the University of Texas at Austin for its financial support of the Advanced School which made these lectures possible. We also should not finish without thanking Barbara Melton for her diligent typing of this manuscript.

William C. Schieve
Acting Director
Center for Statistical
Mechanics and Thermo-
dynamics

CONTENTS

ENTROPY AND DISSIPATIVE STRUCTURE

Ilya Prigogine
Universite Libre de Bruxelles
Belgique
and
University of Texas at Austin

(Text prepared by A. Babloyantz)

CHAPTER I. INTRODUCTION

These lectures are devoted to the problem of structure formation in systems beyond thermodynamic equilibrium. Classical thermodynamics implies that macroscopic structures may arise at equilibrium beyond the phase transition point. A crystal is an example. Once such a structure is formed, it is maintained without any exchange of energy and matter with its surroundings. We shall show that beyond thermodynamic equilibrium new types of critical phenomenon, such as structure formation may occur. These structures are maintained only through exchange of energy (and in some cases of matter) with the outside world. They are "dissipative structures" [1].

Consider a thermodiffusion cell in which a gradient of concentration is maintained by a flow of energy. This exchange of heat leads to an entropy lowering, hence, to an increase of organization. However, the increase occurs gradually with the increase of the gradient of temperature.

The so-called "Benard problem" of classical hydrodynamics is an example of discontinuous changes in structure due to dissipative processes. If a horizontal fluid layer is heated from below, maintaining an "adverse gradient" (to $\vec{g}$) of temperature, for small values of this gradient the fluid remains at rest. But for a critical value of the gradient, there is an abrupt onset of convection, leading to the well known "Benard cell" problem [2].

Dissipative structures are also possible in chemical systems. Let us consider a sequence of reactions such as

$$A \rightleftarrows [X, Y, Z] \rightleftarrows F$$

where the concentrations of the initial and final components A and F are maintained constant. X, Y, Z are intermediate components. The parameter which expresses the thermodynamic constraint may be taken as the ratio of A and F. If this ratio is taken in accordance with the law of mass action, the system will evolve to equilibrium.

If the scheme of reactions is non-linear, there may be different time-independent solutions of the kinetic equations describing the system

$$\frac{dX}{dt} = 0 \qquad \frac{dY}{dt} = 0 \qquad \frac{dZ}{dt} = 0 ,$$

all satisfying the physical condition that the concentrations must be real and positive quantities.

For given values of A and F one of these solutions contains, as a special case, the equilibrium solution corresponding to the minimum of free energy. Will this solution be stable for large deviation from equilibrium?

The occurences of dissipative structures is associated with the fact that the continuation of the equilibrium branch of the kinetic equations becomes unstable and is replaced by another branch. Thus dissipative structures may occur only in non-linear systems for which more than one solution of the kinetic equations exists.

In addition, dissipative structures will occur only at a finite distance from thermodynamic equilibrium as the stability of thermodynamic solution must extend over at least some non equilibrium region.

The preceding remarks lead us to the study of stability properties far beyond equilibrium states.

CHAPTER II. STABILITY OF THERMODYNAMIC SYSTEMS

A. Fluctuations and Balance Equation for Entropy

The structure beyond an instability point in a macroscopic system originates in a fluctuation. Far from the point of instability a fluctuation is followed by a response which brings the system back to the unperturbed state. On the contrary at the point of formation of a new structure the fluctuations are amplified. There is a well known method; normal modes analysis, for the study of these instabilities. However, we want to deduce an independent stability theory based on thermodynamics of irreversible processes and deduce the same information as in normal mode analysis. To this end we have to build a generalized thermodynamic theory of stability which also includes a macroscopic theory of fluctuations.

The starting point is the basic Einstein formula for an isolated system [3]. The probability, P_r, of a fluctuation to some state far from equilibrium is

$$P_r \sim \exp \frac{\Delta S}{k} \quad , \tag{2.1}$$

where ΔS is the change of entropy associated with fluctuations. For small fluctuations (2.1) gives

$$P_r \sim \exp \frac{1}{2k}\, \delta^2 S \quad . \tag{2.2}$$

Here $\delta^2 S$ is the curvature of the entropy surface at the equilibrium state. Assuming local equilibrium (see below) it can be shown that (2.2) remains valid even for fluctuations in non-equilibrium conditions. This has been substantiated recently by detailed calculations based on stochastic models [4,5]. It has been shown that the generalized Einstein formula (where the equilibrium values are replaced by steady state values) remains valid whenever there exists a separation of the time scales between the fluctuating system and the "outside world". Such separation is always implied when well defined boundary conditions are prescribed for the macroscopic systems. Now if we derive a balance equation for $\delta^2 S$ and study its evolution in time, we establish a link between fluctuations and the stability theory.

We shall assume for the remainder of the lectures the validity of the "local equilibrium" assumption. Therefore, the local entropy per unit mass s, is the same function of the local macroscopic variables as given by Gibbs' law at equilibrium [6]:

$$s = s(e, v, N_\gamma)$$

with $\left(\frac{\partial s}{\partial e}\right)_{v,N_\gamma} = T^{-1}$; $\left(\frac{\partial s}{\partial v}\right)_{e,N_\gamma} = pT^{-1}$; $\left(\frac{\partial s}{\partial N_\gamma}\right)_{e,v,N_\gamma} = -\mu_\gamma T^{-1}$

$$\left(N_\gamma = \frac{m_\gamma}{m} \; ; \; \sum_\gamma N_\gamma = 1; \; \gamma = 1,\ldots,c\right) . \qquad (2.3)$$

The local equilibrium assumption implies that the effect of collisions must be sufficiently dominant to exclude large deviations from statistical equilibrium ; an assumption that cannot be maintained for rarefied gases. The balance equation for entropy is [7]:

$$dS = d_e S + d_i S \qquad (2.4)$$

where $d_e S$ denotes the contribution of the outside world and $d_i S$ denotes the entropy production due to the irreversible processes inside the system. The second law of thermodynamics postulates that

$$d_i S \geq 0 . \qquad (2.5)$$

The equality sign corresponds to equilibrium situations where entropy production vanishes.

Now assuming local equilibrium the explicit form of entropy production can be obtained by using the balance equations for mass, momentum, and energy. Then one gets a bilinear expression for the entropy production per unit time [8]:

$$P = \frac{d_i S}{dt} = \int_V dV \sum_\alpha J_\alpha X_\alpha \geq 0 . \qquad (2.6)$$

J_α are the flows (or rates) of the irreversible processes and X_α the corresponding forces.

B. Classical Stability Theory of Thermodynamic Equilibrium

The Gibbs-Duhem stability criterion is a formulation of the second law, valid for closed systems at uniform pressure and temperature [6]:

$$\delta E + p\delta V - T\delta S \geq 0 \quad . \tag{2.7}$$

When this inequality is satisfied, any macroscopic deviation from equilibrium is impossible as inconsistent with a positive entropy production.

For infinitesimal perturbations this gives:

$$(\delta^2 S)_{eq} < 0 \; . \tag{2.8}$$

This kind of approach cannot be extended to the study of non-equilibrium situations due to the lack of thermodynamic potentials.

A more general approach to equilibrium stability valid for all boundary conditions compatible with the maintainance of equilibrium will now be presented.

C. Equilibrium Stability Theory and Entropy Balance Equation

Let us combine (2.4) and (2.6). We get (for more details see ref. 9):

$$P[S] = \int_V dV \sum_\alpha J_\alpha X_\alpha = \partial_t S + \Phi\,[S] \geq 0 \quad . \tag{2.9}$$

$\Phi\,[S]$ is the entropy flow through the surface of the system. Now we separate in the r.h.s. of (2.9) terms of first and second order. The entropy production is a quantity of second order with respect to the deviations from equilibrium. We get

$$\left[\frac{\partial}{\partial t}\,(\Delta S)\right]^{(1)}_{eq} = -\,\Phi\,[S]_{eq} \quad , \tag{2.10}$$

and

$$\frac{1}{2}\,[\frac{\partial}{\partial t}(\Delta S)]^{(2)}_{eq} = P[S] - \Delta\,\Phi\,[S]\,. \qquad (2.11)$$

Eq. (2.10) is a generalized equilibrium condition.

By taking appropriate boundary conditions (2.11) reduces to

$$\frac{1}{2}\,[\partial_t(\Delta S)]^{(2)}_{eq} = P[S] \geq 0\,. \qquad (2.12)$$

According to the Gibbs-Duhem definition of stability, if no perturbation can satisfy the inequality, (2.12) the system will remain in equilibrium. The stability condition is therefore:

$$\frac{1}{2}(\delta^2 S)_{eq} = \int_0^t P[S]dt = \int_0^t d_i S = \Delta_i S < 0. \qquad (2.13)$$

Let us look for the explicit form of the equilibrium stability conditions. It can be shown [see ref. 9] that

$$\delta^2 s = -\frac{1}{T}\,[\frac{C_V}{T}(\delta T)^2 + \frac{\rho}{\chi}\,(\delta v)^2_{N_\gamma} + \sum_{\gamma\gamma'} \mu_{\gamma\gamma'}\delta N_\gamma \delta N_{\gamma'}] \qquad (2.14)$$

with

$$\chi = -\frac{1}{v}(\frac{\partial v}{\partial p})_{T,N_\gamma}\;;\quad (\delta_v)_{N_\gamma} = (\frac{\partial v}{\partial T})_{p,N_\gamma}\,\delta T + (\frac{\partial v}{\partial p})_{T,N_\gamma}\,\delta_p\,; \qquad (2.15)$$

$$\mu_{\gamma\gamma'} = (\frac{\partial \mu_\gamma}{\partial N_{\gamma'}})_{T,p,(N_\gamma)}$$

and

$$(\delta^2 S)_{eq} = \int_V dV\, \rho_{eq}[\delta^2 s]_{eq} < 0\,, \qquad (2.16)$$

this inequality must be verified for arbitrary perturbations. It implies:

$$[\delta^2 s]_{eq} < 0 \qquad (2.17)$$

therefore

$$C_V > 0;\quad \chi > 0;\quad \Sigma\mu_{\gamma\gamma'}x_\gamma x_{\gamma'} > 0\,. \qquad (2.18)$$

These are the classical conditions of equilibrium stability.

D. Stability of Non-Equilibrium States

This study shall be limited to small perturbations with respect to some reference state (the details of this section can be found in reference 8). The entropy production is no longer a second order quantity, so we cannot split the second law as in preceding sections. However we shall take that local equilibrium assumption subsists and equilibrium is stable. Thus inequalities (2.17) and (2.18) remain valid. Therefore, we assume

$$\delta^2 s < 0 \tag{2.19}$$

even for states far from thermodynamic equilibrium. This property suggests an approach to thermodynamic stability which is closely related to the ideas underlying Ljapounov's theory.* We consider $(-\delta^2 s)$ as a Ljapounov function. According to classical stability theory, if

$$\dot{\delta^2 s} > 0 \quad \text{and} \quad \delta^2 s < 0 \ , \tag{2.20}$$

are valid for all times t, the reference state is stable

More explicitly

$$\frac{1}{2}\, \dot{\delta^2 s} = \delta T^{-1} \delta \dot{e} + \delta(pT^{-1}) \delta \dot{v} - \sum_\gamma \delta(\mu_\gamma T^{-1}) \delta \dot{N}_\gamma \geq 0 \ , \tag{2.21}$$

the "dot" denotes a time derivative in the space of the independent variables e, v, N_γ.

Since a local formulation of stability theory is not appropriate for a description of systems submitted to given boundary conditions, we need a global formulation.

Using the local formulation one can show [8,9] that a global stability condition is:

*For a discussion of this see La Salle, J. and Lefschetz, S., Stability by Ljapounov's Direct Method, Acad. Press Inc. (New York, 1961).

$$\delta^2 S < 0; \quad \partial_t \delta^2 S \geq 0 \quad . \tag{2.22}$$

To investigate the stability of a given non-equilibrium state by means of this criterion it remains to establish the explicit form of the entropy-balance equation associated with small perturbations. We shall use the balance equations for mass, momentum, energy and assuming the system is in mechanical equilibrium,we get:

$$\frac{1}{2} P[\delta^2 S] = \int_V dV \sum_\alpha \delta J_\alpha \delta X_\alpha \geq 0 \quad . \tag{2.23}$$

We shall call this important quantity the excess entropy production. The similarity in structure with the entropy production (2.6) is striking. However, while the very formulation of the second law of thermodynamics prescribes a positive value of P, the sign of the excess entropy production depends on the kinetic laws relating the fluxes J_α and the generalized forces X_α

An especially interesting case is the one of chemical reactions, where the laws between reaction rates and chemical affinities are generally non linear. The stability condition (2.23) becomes:

$$\sum_\rho \delta\omega_\rho \delta(A_\rho T^{-1}) \geq 0 \quad , \tag{2.24}$$

where ω_ρ are the chemical reaction rates, and A_ρ the corresponding affinities.

CHAPTER 3. CHEMICAL EXAMPLES

We shall study the behavior of chemical reactions in open systems, far from equilibrium, and look for possible new effects beyond instability. We shall show that the dissipation may give rise to space and time order. We shall consider systems exhibiting the following behavior:

a) oscillations occur on the thermodynamic branch, but no instability appears (Lotka - Volterra Model),

b) oscillations occur after an instability point (on a new non thermodynamic branch) ,

c) there is structure formation beyond an instability point.

In this section we combine the results of the thermodynamic analysis outlined in previous sections and derive the behavior of a single normal mode around the steady state.

We take fluctuations of the form

$$\begin{aligned} \delta\omega_\rho &= \delta\omega^o_\rho \quad \exp[(\omega_1 + i\omega_2)t] \\ \delta A_\rho &= \delta A^o_\rho \quad \exp[(\omega_1 + i\omega_2)t] \quad . \end{aligned} \tag{3.1}$$

Using (2.24) and (3.1) it can be shown that [8].

$$\frac{1}{2} \int dV\{\omega_1 \sum_\rho(\delta\omega_\rho^*\delta A_\rho + \delta\omega_\rho\delta A^*_\rho) + i\omega_2 \sum_\rho(\delta\omega^*_\rho\delta A_\rho - \delta\omega_\rho\delta A^*_\rho)\}$$

$$= \omega_1 \delta_m P + i\omega_2\delta_m\Pi \leq 0 \quad . \tag{3.2}$$

$\delta_m P$ may be associated with the approach to the steady state, and $\delta_m\Pi$ with rotation around this state. It can easily be seen that the stability condition may be written as [9].

$$\omega_1\delta_m P \leq 0 \quad . \tag{3.3a}$$

At the same time one can obtain a criterion for the onset of oscillations

$$\omega_2\delta_m\Pi \leq 0 \quad . \tag{3.3b}$$

A. Oscillation On Thermodynamic Branch

Let us consider the following scheme of reactions which contains an autocatalytic step [8,10]

$$A + X \xrightarrow{k_1} 2X$$

$$X + Y \xrightarrow{k_2} 2Y \qquad (3.4)$$

$$Y \xrightarrow{k_3} E$$

where the values of initial and final products A and E are held constant.

The corresponding kinetic equations are:

$$\frac{dX}{dt} = k_1 AX - k_2 XY$$

$$\frac{dY}{dt} = k_2 XY - k_3 Y \qquad (3.5)$$

They admit a single non zero steady state solution.

We perform a standard normal mode analysis around this stationary state. The dispersion equation indicates that small fluctuations around the steady state are periodic with frequency

$$\omega = i(k_1 \; k_3 \; A)^{1/2} \, . \qquad (3.6)$$

Moreover, a Thermodynamic analysis shows that [8]

$$\delta_m P = 0 \qquad \text{and} \qquad \delta_m \Pi \geq 0 \, . \qquad (3.7)$$

Consequently, one may conclude that the system is in a state of "marginal stability"; fluctuations can neither decay nor grow.

Upon integrating the kinetic equations one gets:

$$X^{-1} \, e^{X} = e^{K} \, Y \, e^{-Y} \qquad (3.8)$$

where K is an arbitrary constant, and is fixed by the initial conditions. A particular value of K gives a corresponding closed curve in the X,Y plane; therefore, even large fluctuations around the steady state are periodic. This time however the periods depend on the initial conditions.

B. Oscillation Past An Instability: Limit Cycles

Let us study a case of instability for which the excess entropy production first vanishes and then changes sign for a finite value of the overall affinity. The thermodynamic branch then becomes stable.

We consider the following scheme of reactions [8,10,11]

$$\begin{aligned} A &\underset{k}{\overset{1}{\rightleftarrows}} X \\ 2X + Y &\underset{k}{\overset{1}{\rightleftarrows}} 3X \\ B + X &\underset{k}{\overset{1}{\rightleftarrows}} Y + D \\ X &\underset{k}{\overset{1}{\rightleftarrows}} E \end{aligned} \tag{3.9}$$

where A, B, D, E are held constant. The kinetic equations are (if we take k=0):

$$\begin{aligned} \frac{dX}{dt} &= A + X^2Y - BX - X \\ \frac{dY}{dt} &= BX - X^2Y \; . \end{aligned} \tag{3.10}$$

The steady state values of X and Y are

$$X_o = A \qquad\qquad Y_o = B/A \tag{3.11}$$

the usual normal mode analysis considering again only homogeneous perturbations yields

$$\omega^2 + (X_o^2 + B + 1 - 2X_oY_o)\omega + X_o^2 = 0. \tag{3.12}$$

The value of B for which the coefficient of ω in (3.12) vanishes corresponds to a transition point. Beyond this point ($B > B_c$) the system becomes unstable. We find

$$B_c = 1 + A^2 . \tag{3.13}$$

In addition, Poincaré has shown that differential equations of the form (3.10) beyond a point of instability such as given by (3.13) may admit solutions represented by closed curves in the phase plane (hence periodic solutions) which are such that another trajectory which is also closed is necessarily at a finite distance from the former: these are the limit cycles. The fundamental importance of limit cycles is to represent self sustained oscillatory motion in non linear, non conservative systems. This can be seen on Figure 1.

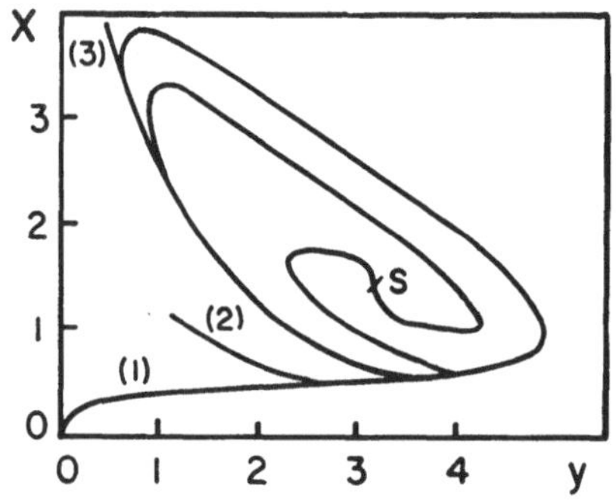

FIGURE 1

Figure 1 shows the typical behavior of the system in a region where a limit cycle is observed. Whatever the initial conditions (1,2,3) the system reaches the same periodic behavior.

Let us evaluate $\delta_m P$ around the steady state. We obtain [ref. 8]

$$\delta_m P = \frac{1}{AB} [B(1-B)\delta X\delta X^* + A^4 \delta Y \delta Y^*] . \tag{3.14}$$

$B\delta X\delta X^*$ is due to the autocatalytic step and is the "dangerous" contribution. For $B \leq 1$ (3.14) is positive definite and the system is stable.

It can be shown that for $B = B_c$, $\delta_m P$ vanishes and $\delta_m \Pi \geq 0$. The thermodynamic criterion therefore gives the same information in a concise form, as the kinetic methods.

C. Symmetry Breaking Instabilities

So far we have discussed the problem of stability in chemical systems with respect to homogeneous fluctuations. Both the unperturbed and perturbed systems were assumed homogeneous. The problem of stability with respect to diffusion will now be considered.

Let us return to the scheme (3.9). In the kinetic equations (3.10), we have to add $D_X \frac{\partial^2 X}{\partial^2 r^2}$ and $D_Y \frac{\partial^2 Y}{\partial r^2}$ in order to take account of diffusion. We take a perturbation of the form

$$X(t) - X_o = x \quad \exp\,(\omega t + \frac{ir}{\lambda})$$

$$Y(t) - Y_o = y \quad \exp\,(\omega t + \frac{ir}{\lambda})$$

It can then be shown that the homogeneous stationary state becomes unstable (for the details of this section see refs. [8,12]), for a critical wave length λ_c, and B_c.

$$\lambda_c^2 = \frac{1}{A}\,(D_x D_y)^{1/2}$$

$$B_c = [1 + A(\frac{D_x}{D_y})^{1/2}]^2 \qquad (3.15)$$

Since the computational problem involved in the study of the time evolution beyond the instability is rather involved, we shall consider rather than disturbances of arbitrary wavelength a two box model. In this system the initial and final products are distributed homogeneously in the system whereas X and Y may diffuse freely between the two parts. We now have four equations:

$$\frac{dX_1}{dt} = A + X_1^2 Y_1 - BX_1 - X_1 + D_x(X_2 - X_1)$$

$$\frac{dX_2}{dt} = A + X_2^2 Y_2 - BX_2 - X_2 + D_x(X_1 - X_2)$$

$$\frac{dY_1}{dt} = BX_1 - X_1^2 Y_1 + D_y(Y_2 - Y_1)$$

$$\frac{dY_2}{dt} = BX_2 - X_2^2 Y_2 + D_y(Y_1 - Y_2) . \qquad (3.16)$$

The time independent homogeneous solution is

$$X_i = A \qquad Y_i = B/A \qquad (i = 1,2) \qquad (3.17)$$

We make the following choice of numerical values for the diffusion coefficient of X and for A

$$D_x = 1 \qquad A = 2 .$$

There remains two arbitrary parameters whose values determine the properties of the steady states. The steady state equations are:

$$3X_2^5 - 30X_2^4 + [96+2D_y(B+3)]X_2^3 - [96+12D_y(B+3)]X_2^2 + 16D_y(6+B)X_2 - 96D_y = 0$$

$$X_1 = 4 - X_2$$

$$Y_2 = B(8X_2^2 - 4D_y - 16X_2 - X_2^3)/[8X_2^3 - 3X_2^2 - (D_y+8) + 8D_y(X_2-2) - X_2^4]$$

$$Y_1 = Y_2 + (X_2^2 Y_2 - BX_2)/D_y . \qquad (3.18)$$

This system has two types of solutions: the homogeneous steady state solution given by (3.17) and an inhomogeneous solution which because of the symmetry of the model may be written in two equivalent ways.

$$X_1 > X_2, \qquad Y_1 < Y_2 \qquad (3.19)$$

or

$$X_1 < X_2, \qquad Y_1 > Y_2$$

The stability analysis of the steady state solutions of (3.18) has been performed both for homogeneous and inhomogeneous fluctuations. One finds that the homogeneous state is unstable with respect to homogeneous perturbation when

$$B > 5 \qquad (3.20)$$

and with respect to inhomogeneous perturbations when

$$B > B_c = \frac{1}{2D_y}(6D_y + 12) \quad . \qquad (3.21)$$

These results combined with those obtained for the inhomogeneous steady state are shown in Figure 2.

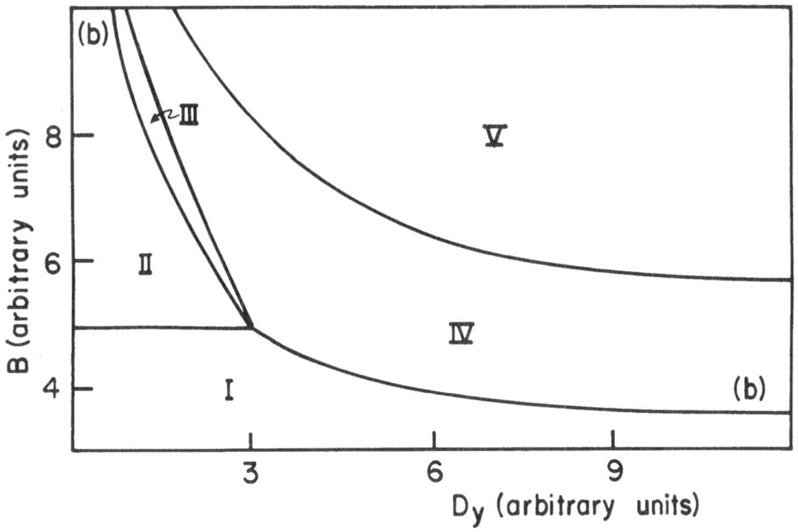

FIGURE 2

Conditions (3.20) and (3.21) define a domain I where only the homogeneous steady state exists and is stable. This state becomes unstable with respect to homogeneous perturbations in II and with respect to inhomogeneous perturbations beyond the curve (b). In regions II, III, V no time independent, stable state exists, but in region IV the inhomogeneous steady state is stable. This region, therefore, corresponds to what we call a dissipative structure. It is particularly interesting to investigate how such an inhomogeneous state is reached. A typical result is reported in Figure 3.

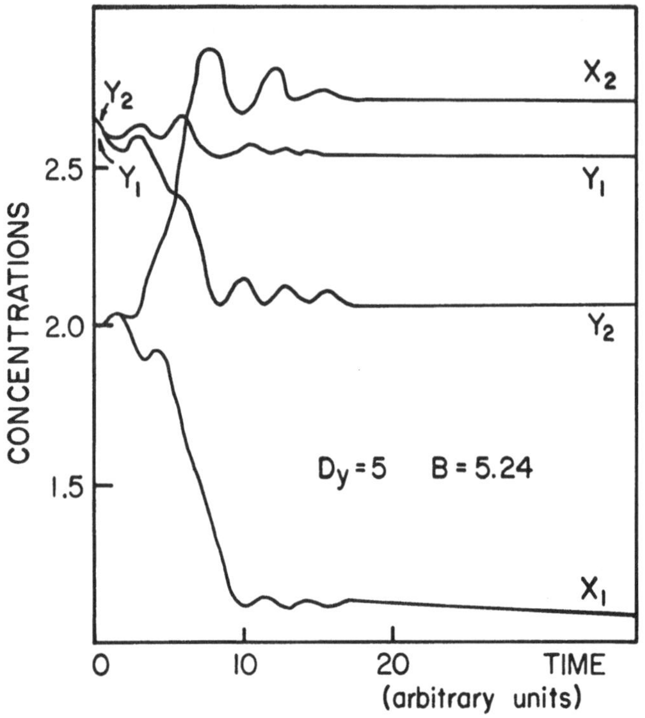

FIGURE 3

The homogeneous state corresponding to $X_1=X_2=2$ and $Y_1=Y_2=2.62$ is destroyed by a small fluctuation $Y_2-Y_1=0.04$. It is clearly seen how the initial perturbation is magnified until the inhomogeneous steady state is reached. The configuration (3.19) chosen by the system depends crucially on the nature of the initial perturbation. The thermodynamic analysis yields

$$\delta_X P = (1-B)\frac{X^2}{A} + \frac{A^3}{B} Y^2 + \frac{D_X}{\lambda^2 A} x^2 + \frac{D_Y}{\lambda^2 B} y^2 . \qquad (3.22)$$

Again, in agreement with our general discussion, a negative term, $-\frac{B}{A} X^2$, due to the auto catalytic action of X, appears. This is the "dangerous contribution".

The explicit contribution of diffusion is positive and proportional to $\frac{D}{\lambda^2}$. Therefore, if there is an instability, increasing values of D must give rise to increasing values of the critical wavelength. If not, the contribution of diffusion to (3.22) would become dominant and $\delta_X P$ would be always positive. Diffusion has a second role - the manifold of perturbations which we may introduce into (3.22) is now increased by the consideration of inhomogeneous systems.

Dissipative structures have been observed experimentally. We will not describe them here but the reader is referred to papers by Zhabotinsky [12], Busse [13] and Herschkowitz-Kaufman [14].

References

[1] I. PRIGOGINE: "Structure, Dissipation and Life" in Theoretical Physics and Biology, ed. M. Marois (Amsterdam: North Holland Publishing Co., 1969)
I. PRIGOGINE: "Dissipative Structures in Biological Systems" Second International Conference on 'Theoretical Physics and Biology', Versailles (July 1969)

[2] S. CHANDRASEKHAR: Hydrodynamic and Hydromagnetic Stability (Oxford: Clarendon Press, 1961)

[3] L. D. LANDAU AND E. M. LIFSHITZ: Statistical Physics (New York: Addison-Wesley, 2nd ed. 1969) Ch. XII

[4] G. NICOLIS AND A. BABLOYANTZ: J. Chem. Phys. 51, 2632 (1969)

[5] A. BABLOYANTZ AND G. NICOLIS: J. Stat. Phys. 1,

[6] I. PRIGOGINE AND R. DEFAY: Thermodynamique Chimique Editions Desoer, Liege (1950)

[7] I. PRIGOGINE: Introduction to Thermodyanmics of Irreversible Processes, 3rd edition (New York: Interscience, John Wiley and Sons, 1967)

[8] P. GLANSDORFF AND I. PRIGOGINE: Generalized Thermodynamics (in press) Wiley, Interscience (1970)

[9] P. GLANSDORFF AND I. PRIGOGINE: Physica 46, 344 (1970)

[10] I. PRIGOGINE AND G.NICOLIS: J. Chem. Phys. 46, 3542 (1967
G. NICOLIS: Adv. Chem. Phys., to appear 1970
R. LEFEVER, G. NICOLIS, AND I. PRIGOGINE; J. Chem. Phys. 47, 1045 (1967)

[11] I. PRIGOGINE AND R. LEFEVER: J. Chem. Phys. 48, 1605 (1968)
R. LEFEVER: J. Chem. Phys. 49, 4977 (1968)
Bull. Cl. Sci. Acad. Roy. Belg. 54, 712 (1968)

[12] A. M. ZHABOTINSKY: Dokl. Acad. Nauk. U.S.S.R. 152, 392 (1964)
Biofizika 9, 306 (1964)

[13] H. BUSSE: J. Phys. Chem. 73, 750 (1969)

[14] M. HERSCHKOWITZ-KAUFMAN: C.R. Acad. Sc. Paris 270, 1049 (1970)

PHASE TRANSITIONS

Zevi W. Salsburg

(Text prepared from handwritten and tape recorded lecture notes of the late author by M. G. Velarde and R. W. Gibberd)

FOREWORD

An interesting problem in the field of equilibrium statistical mechanics is the explanation of phase transitions and the calculation of the properties of such transitions. In recent years these problems have received a good deal of attention and the efforts have been rewarded by a number of significant advances and discoveries, although no complete theory is yet available. One can hardly hope to describe in any detail all those advances in a set of four lectures,as a detailed treatment of the theory of "critical phenomena" could form the basis of an entire semester course, and so we must settle for some compromise. I will first try to give what could pass as an outline for a course on phase transitions, emphasizing a few special points. We will then turn to one special topic, namely the question of the existence or non-existence of phase transitions and long-range order in 1, 2 and 3 dimensional space. This topic has been selected arbitrarily,but hopefully the general outline will put it in a proper perspective.

CHAPTER I. CLASSICAL DESCRIPTION OF PHASE TRANSITIONS

A. Introduction

I will start by assuming familiarity with the general thermodynamic treatment of phase transitions which leads to the equality of temperature, pressure and chemical potential across phase boundaries and the formulation of the Gibbs phase rule as, for example given by Landau and Lifshitz [1]. However, to establish notation, I will summarize some of the basic concepts for the special case of a single component system. We focus our attention on the Gibbs free energy, G(T,P,N), as a function of the temperature T, pressure P, and number of particles N. G is assumed to be a continuous function for the entire accessible region of thermodynamic states. For each homogeneous phase (region in T,P,N space where G is differentiable) we have the fundamental equation

$$dG = -SdT + Vdp + \mu dN , \qquad (1.1)$$

and the Gibbs-Duhem relation [obtained by using (1.1) with the relation $dG = Nd\mu + \mu dN$]

$$SdT - VdP + Nd\mu = 0 \quad . \qquad (1.2)$$

When two phases (denoted by ' and ") are in equilibrium, thermodynamics requires that

$$T' = T'', \qquad P' = P'', \qquad \mu' = \mu'' . \qquad (1.3)$$

For a fixed value of N one can picture G(T,P,N) as a surface over the T,P plane. This surface will be analytic except along certain lines and isolated points called the phase boundaries. The projection of these boundaries into the T,P plane gives a phase diagram as shown in Figure 1 where C denotes the critical point and t denotes the triple point.

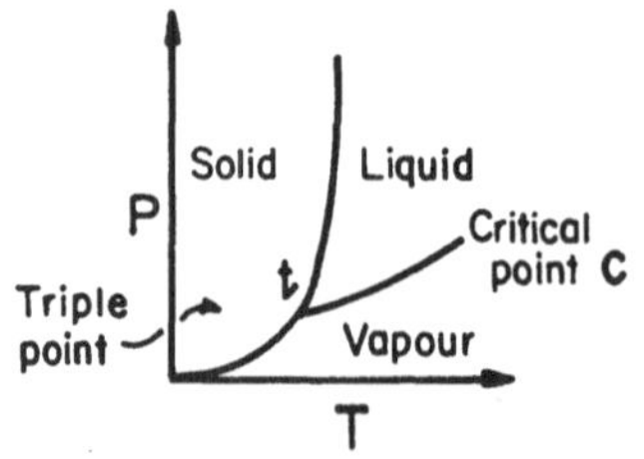

FIGURE 1. Phase diagram of a pure fluid in the temperature-pressure plane

In a multicomponent system the classification of singular points such as the critical point becomes a complex problem. As an example consider a binary mixture and the three independent intensive variables

$$T,P,X \qquad X = \frac{N_2}{N_1+N_2} \quad .$$

The phase boundaries must now be pictured as surfaces in T,P,X space. For the liquid-gas transition one finds the qualitative situation as depicted in Figure 2.

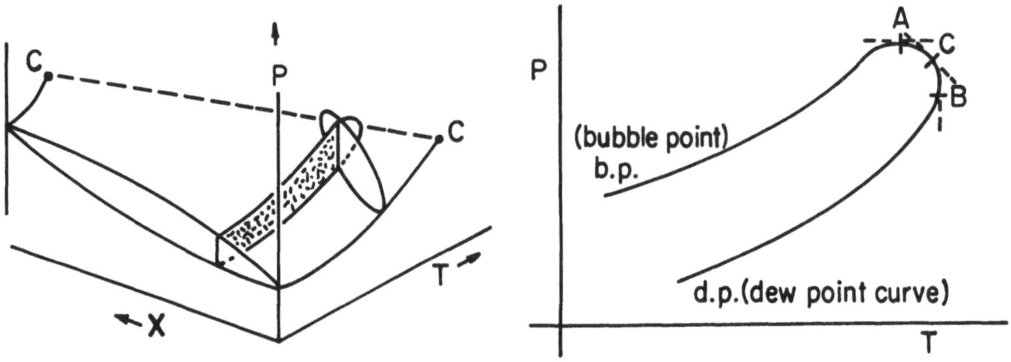

FIGURE 2a. Liquid-gas transition 2b. A T-P cross-section

In looking at a typical T-P cross-section (represented by the shaded area in Figure 2a, we might encounter the situation shown in Figure 2b. In this diagram A is the point of maximum pressure on the coexistence curve, B is the point of maximum temperature on the coexistence curve and C, the critical or plait point represents the state in which the two phases have become identical. It is the point at which the first singularity develops in the Gibbs-free energy function.

B. Ehrenfest Classification

The first attempt to describe in a unified way phase transition phenomena was made by Ehrenfest [2]. As we will see later on the Ehrenfest classification is not entirely appropriate but it provides some insight into the different types of phase transitions that we may expect to occur.

The Ehrenfest classification of phase transitions is based upon the properties of a thermodynamic potential function such as G(T,P,N). A transition is classified as being of 1st, 2nd, 3rd, etc... order. An n-th order transition is one in which the first (n-1) derivatives of G are continuous (and so finite) across the phase transition, while at least one of the n-th derivatives is discontinuous. A typical example of a first order phase transition (n=1) would be where the entropy, $S = -(\frac{\partial G}{\partial T})_{P,N}$, is discontinuous. This is characteristic of the commonly known transition such as melting and condensation. The phase diagram boundaries for a single component system are described by the Clapeyron equations which can be obtained as follows.

For two phases (' and ") in equilibrium one can write a Gibbs-Duhem relation for each phase

$$S'dT - V'dP + N'd\mu = 0 = S''dT - V''dP + N''d\mu \ . \qquad (1.4)$$

From these two equations we solve for $\frac{dP}{dT}$ and $\frac{d\mu}{dT}$

$$\frac{dP}{dT} = \frac{s'-s''}{v'-v''} \tag{1.5a}$$

$$\frac{d\mu}{dT} = \frac{v's'' - v''s'}{v' - v''} \quad , \tag{1.5b}$$

where v = V/N and s = S/N denote the specific volume and entropy per particle respectively.

While first order transitions are clearly realized and recognized in a variety of physical systems there has been considerable controversy about whether or not real second order phase-transitions exist in nature. Moreover, a good deal of discussion has centered about a physical explanation for second order transitions. We will return to this point in the outline of Landau's theory of second order transitions.

Typically one looks for a situation in which S is continuous across the transition but, for example, the heat capacity

$$C_p = T\left(\frac{\partial S}{\partial T}\right)_P = -T\left(\frac{\partial^2 G}{\partial T^2}\right)_P \quad , \tag{1.6}$$

would have a discontinuity.

The phase boundary or coexistence line for such transitions can be described by the Ehrenfest relations. We start with

$$V' - V'' = \Delta V = 0 \quad , \tag{1.7a}$$

and

$$S' - S'' = \Delta S = 0 \quad , \tag{1.7b}$$

and differentiate these relations along the coexistence curve, P = P(T),

$$\frac{\partial \Delta V}{\partial T} + \frac{\partial \Delta V}{\partial P}\frac{dP}{dT} = 0 \tag{1.8a}$$

$$\frac{\partial \Delta S}{\partial T} + \frac{\partial \Delta S}{\partial P}\frac{dP}{dT} = 0 \quad , \tag{1.8b}$$

or

$$\Delta\left(\frac{\partial V}{\partial T}\right)_p + \Delta\left(\frac{\partial V}{\partial P}\right)_T \frac{dP}{dT} = 0 \qquad (1.9a)$$

$$\frac{\Delta C_p}{T} - \Delta\left(\frac{\partial V}{\partial T}\right)_p \frac{dP}{dT} = 0 \quad , \qquad (1.9b)$$

From these two equations one can find $\left(\frac{dP}{dT}\right)$ and also one of many relations among the discontinuities

$$\frac{dP}{dT} = \frac{\Delta C_p}{T\Delta\left(\frac{\partial V}{\partial T}\right)_p} = -\frac{\Delta\left(\frac{\partial V}{\partial T}\right)_p}{\Delta\left(\frac{\partial V}{\partial p}\right)_T} \; . \qquad (1.10)$$

Recently Grindlay [3] has described a physical case where such second order phase transitions are realized, namely a superconductor subject to varying pressure, in zero external magnetic field.

C. Landau's Theory of Second-Order Transitions

In exploring the question of classifying phase transitions, Landau [1] distinguished two different cases: (a) transitions in which the two different phases were in different physical "states" e.g. a f.c.c. structure in equilibrium with a b.c.c. structure (ordinary phase transition or first order phase transition); (b) transitions in which the "state" of the system does not change at the transition point but a symmetry element appears or disappears as the system passes through the transition (second-order phase transitions or Curie or λ-points).

The Landau theory then proceeds along these two lines: (a) a loss of an element of symmetry is always accompanied by a discontinuity in the heat capacity (this led Landau to doubt the existence of higher-order transitions, since he could not imagine any other transition mechanism, other than a change in state or a change in symmetry); (b) given a physical model with a certain symmetry, what changes of symmetry are possible in a second-order phase transition? This involves some general group theory arguments. The number of possibilities are greatly restricted by the conditions of thermodynamic stability.

The Landau theory postulates the existence of an order parameter, η, which vanishes ($\eta=0$) upon the increase in symmetry. That is $\eta=0$ in the more symmetric state and $\eta=\eta(\tau)$ in the less symmetric one. The important fact is then that the symmetry element in a crystal for example is changed only when η becomes exactly $\eta=0$; any non-zero order parameter, however small, brings about the same symmetry as that of a completely ordered crystal. If, as the temperature is increased the order parameter becomes zero discontinuously from some finite value then the transition from an ordered to a disordered crystal is a first-order transition. If the order parameter goes to zero in a continuous way, i.e. without finite jump, then we have a second-order phase transition.

For any given value of η we can formulate the Gibbs free energy $\hat{G}$ under the constraint of fixed η

$$\hat{G} = \hat{G}(T,P,\eta) \quad . \tag{1.11}$$

For a given thermodynamic state (T,P) the physical value of η is determined by minimizing the Gibbs free energy, the necessary condition is

$$\left(\frac{\partial \hat{G}}{\partial \eta}\right)_{\eta=\eta^*(T,P)} = 0 \quad , \tag{1.12}$$

and

$$G(T,P) = \hat{G}(T,P,\eta^*[T,P]) \quad . \tag{1.13}$$

Landau proceeds further with the assumption that in the neighborhood of the transition point $\hat{G}(T,P,\eta)$ can be expanded as follows

$$\hat{G}(T,P,\eta) = G_0 + A(T,P)\eta^2 + C(T,P)\eta^4 + \ldots \quad . \qquad (1.14)$$

From this it follows that $A(T_c,P) = 0$ determines the transition point with $A(T,P) > 0$ above the transition point in the more symmetric phase and < 0 below the transition point in the less symmetric phase

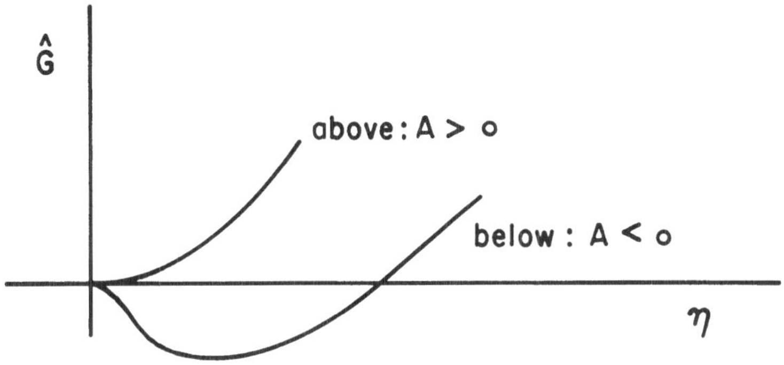

FIGURE 3. A thermodynamic potential as a function of the order parameter

Note that the linear term in (1.14) is identically vanishing as it is assumed that the states for $\eta = 0$ and $\eta \neq 0$ are distinguished by their symmetry elements.

Note also that the cubic term in the expansion of G is identically zero. This comes about by requiring thermodynamic stability (a minimum in $\hat{G}$) at the transition point. In the more general group theory analysis this requirement means that there are no invariants of the third order associated with the lost element of symmetry.

The fourth order term is positive because of the same stability argument. C, being positive for $\eta = 0$, is also positive in the neighborhood of the transition point.

One then has a locus A(P,T) = 0 of second-order phase transition points.

The Landau theory in the original form relied upon Taylor series expansions of the type given in (1.14) and as a result was not applicable in the immediate neighborhood of a transition point. In recent years the theory has been reformulated to avoid such expansions and now has a purely group theoretic formulation [4].

D. Tisza Theory of Phase Transitions

Tisza [5] considered the problem of higher-order phase transitions from a purely macroscopic point of view. Tisza noted that higher order transitions observed in nature were characterized by λ point anomalies in the specific heat rather than discontinuous jumps as predicted by the Ehrenfest and Landau theories. The two types are illustrated in Figure 4:

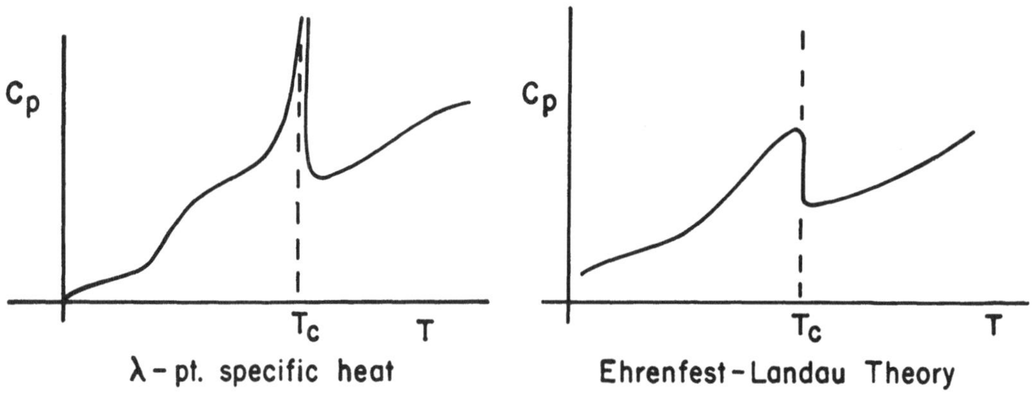

FIGURE 4. Ehrenfest-Landau theory

Moreover, Onsager's exact solution of the Ising model gave a logarithmic singularity in the specific heat [6].

To reformulate the problem Tisza turned to the theory of thermodynamic stability. One starts by considering the internal energy as a function of the extensive state variables. As examples, the energy in a single-component system can be written, E = E(S,V,N) and for a substance in a magnetic field, $E = E(S,V,N,\vec{B})$. We can write in general $E = E(\chi_1,\chi_2,\ldots\chi_n)$.

An expansion of E in powers of $\delta\chi_i = (\chi_i - \chi_i^o)$ around some $\{\chi_i^o\}$ equilibrium state yields

$$E = E^o + \delta E + \delta^2 E + \ldots. \tag{1.15}$$

Thermodynamic stability requires that

$$\delta^2 E > 0 \tag{1.16}$$

Second order phase transitions were then characterized by Tisza as points of neutral stability, i.e. points at which

$$\delta^2 E = 0 \tag{1.17}$$

This theory predicted that C_p, $(\frac{\partial v}{\partial T})$, $(\frac{\partial v}{\partial P})$ approach infinity at the transition points, a result that seemed in agreement with the observed λ-point transitions. For more details the reader may refer to [5].

E. The Divergence in the Specific Heat

Physical reality requires, of course, that we cannot have thermodynamic states for which quantities such as energy or entropy are infinite. It is also clear that the specific heat, C_V for example, cannot be infinite along any finite path in a given phase or else one could integrate $\int C_V dT$ to obtain an infinite energy.

However the question still remains: can C_V (or any other appropriate physical quantity) be infinite only at isolated points (a prediction of Landau's theory) or can one have a phase transition boundary along which $C_V = \infty$? This question was recently treated in an article by Weeler and Griffiths [7]. The result is that a locus of points of infinite C_V is in general incompatible with thermodynamic stability. The argument can be illustrated as follows.

Let us consider a locus of points, $V_0(T)$, in the V-T plane. Define the derivatives of S,V and P along the line $V_0(T)$ as

$$S' = \frac{dS}{dT} \qquad P' = \frac{dP}{dT} \qquad V' = \frac{dV}{dT} \;. \tag{1.18}$$

Then

$$S' = \left(\frac{\partial S}{\partial T}\right)_V + \left(\frac{\partial S}{\partial V}\right)_T V' = \left(\frac{\partial S}{\partial T}\right)_V + \left(\frac{\partial P}{\partial T}\right)_V V' \tag{1.19a}$$

$$P' = \left(\frac{\partial P}{\partial T}\right)_V + \left(\frac{\partial P}{\partial V}\right)_T V' \;. \tag{1.19b}$$

Thus

$$S' - P'V' = \frac{C_V}{T} - \left(\frac{\partial P}{\partial V}\right)_T (V')^2 \;. \tag{1.20}$$

Thermodynamic stability requires that

$$-\left(\frac{\partial P}{\partial V}\right)_T \geq 0 \;. \tag{1.21}$$

Thus we have the basic inequality

$$S' - P'V' \geq \frac{C_V}{T} \;. \tag{1.22}$$

We next integrate this expression along $V_0(T)$ [or at a distance ε to one side] between points a and b

$$\int_a^b S'dT - \int_a^b P'V'dT = S(b) - S(a) - \int_a^b P'V'dT$$

$$= S(b) - S(a) - [P(b)V_0'(b) - P(a)V_0'(a)] + \int_a^b PV_0''(T)dT \geq \int_a^b \frac{C_V}{T}\, dT \;. \tag{1.23}$$

Since S and P are bounded, we conclude that there is a finite upper bound for the integral of C_V/T along any finite locus in the V-T plane and hence C_V can only diverge at a finite number of isolated points.

F. Classification by Broken Symmetry

The Landau theory emphasized the role of symmetry in second-order phase transitions but still attempted to fit the analyis into Ehrenfest's classification scheme. However within the last decade there has been an increasing emphasis of the role of symmetry and group theory in all phase transitions.

R. H. Brout [8] has proposed that phase transitions be classified by their symmetry characteristics. One first notes that in each phase transition the "condensed" phase exhibits a loss of symmetry. The symmetry that is broken may involve a group of discrete transformations or a group of continuous transformations.

In the classification one selects a property of the system, called the response coordinate (e.g. the long-range order parameter), which exhibits the loss of symmetry. In the case of ferromagnetism one could choose the total magnetization $\vec{M}$, whose direction violates the basic isotropy of the system. The details of this type of classification are well explained by Brout [8].

CHAPTER II. STATISTICAL MECHANICS OF PHASE TRANSITIONS

A. Introduction

In principle one would like to start with the partition function (or some equivalent formulation) for one of the statistical mechanical ensembles and deduce the existence of a phase transition and calculate the characteristic parameters such as the condensation temperature (as a function of pressure), the critical parameters etc. Let us start by taking this approach. We consider three commonly used ensembles [9]:

1) Canonical Ensemble

$$e^{-\beta A(N,V,T)} = Q(N,V,T) = \sum_{\substack{i \\ \{\text{accessible} \\ \text{states}\}}} e^{-\beta E_i} \tag{2.1}$$

$$\beta = 1/k_B T \quad .$$

In the classical limit we are concerned with the following form

$$e^{-\beta A(N,V,T)} = \frac{1}{h^{3N} N!} \int d\underset{\sim}{P}^N \int d\underset{\sim}{R}^N \, e^{-\beta H_N} \tag{2.2}$$

$$= \frac{1}{N!\Lambda^{3N}} \int d\underset{\sim}{R}^N \, e^{-\beta\Phi_N} , \tag{2.3}$$

where

H_N = the Hamiltonian function of the system,

$\Phi_N = \Phi_N(\vec{R}_1,\ldots,\vec{R}_N)$ = intermolecular potential,

$\Lambda = \left(\frac{h^2}{2\pi m k_B T}\right)^{1/2}$ = thermal de Broglie wave-length,

$A(N,V,T)$ = Helmholtz free energy,

2) Isothermal-Isobaric Ensemble

$$e^{-\beta G(N,P,T)} = Q(N,P,T) = \beta Q\int_0^\infty e^{-\beta PV}\, Q(N,V,T)dV \qquad (2.4)$$

where

P = pressure

G(N,P,T) = Gibbs free energy.

3) Grand-Canonical Ensemble

$$e^{\beta VP\,(\mu,V,T)} = Q(\mu,V,T) = \Xi = \sum_{N=0}^{\infty} Q(N,V,T)\lambda^N, \qquad (2.5)$$

where

μ = chemical potential per particle

$\lambda = e^{\beta\mu}$ = absolute activity.

For finite systems one has the well known result that all the above partition functions are smooth analytical functions of the primary variables.

The question, then, of how one can obtain sharp transitions from statistical mechanics is partially answered by the fact that discontinuities or singularities in the thermodynamic properties can only occur when one considers the limit of an infinitely large system. Of course, any real system does have a finite size, but with $N \cong 10^{23}$ one hopes that the model of an infinite system will be a close approximation to the actual situation encountered in nature.

The basic problem then reduces to the study of the statistical mechanics of infinite sized systems (systems with infinitely many degrees of freedom).

In recent years there have been some direct approaches to this problem, one being the C^* algebra methods introduced by D. Ruelle and N. H. Hugenholtz [10, 11]. A comprehensive book dealing with this technique as applied to Statistical Physics is in preparation by

G. Emch.

The more traditional approach always starts with the statistical mechanics of finite systems and then goes to the thermodynamic limit ($N, V \to \infty$, $N/V = \rho$). More specifically we consider the following limiting processes for each ensemble.

1) Canonical Ensemble

$$\lim_{\substack{N,V \to \infty \\ N/V = \frac{1}{v}}} \left\{\frac{1}{N} \frac{A(NVT)}{k_B T}\right\} = \beta a(T,v) \tag{2.6}$$

2) Isothermal-Isobaric Ensemble

$$\lim_{\substack{N \to \infty \\ p = \text{const}}} \left\{\frac{1}{N} \frac{G(N\ P\ T)}{k_B T}\right\} = \beta g(T,p) \tag{2.7}$$

3) Grand Canonical Ensemble

$$\lim_{\substack{V \to \infty \\ \mu = \text{const.}}} \frac{\{P(\mu VT\}}{k_B T} = \beta p(T,\mu) \tag{2.8}$$

B. Appearance of Phase Transitions

In carrying out a statistical mechanical calculation how does one recognize the existence of a phase transition in the model? This is a question which can be answered in a number of different ways. We will try to give a qualitative indication of a few of these techniques.

1) Coexistence Theories

In most practical calculations one calculates the Gibbs free energy in each phase by two independent models or approximations and then determines the transition point by the equality of the two free energies

$$G_1(T,P) = G_2(T,P) \tag{2.9}$$

For example for the liquid-solid transition one could use a Percus-Yevick [12] equation of state for the liquid and a harmonic vibration model for the solid.

The region where $G_1 = \mathrm{Min}\{G_1, G_2\}$ represents the region where phase 1 is stable and vice versa.

This type of approach carries with it a description of meta-stability in the sense that each function, G_1 for example, can be continued into the region where the other function, G_2, is a minimum.

Ideally one would like to see the coexistence theory arise from one formulation of the partition function. The following artificial example will illustrate how this can occur.

Suppose that the N,P,T ensemble gives the following result for the partition function

$$Q(NPT) = [f_1(P,T)]^N + [f_2(P,T)]^N \quad . \tag{2.10}$$

Then

$$\beta g(P,T) = -\lim_{N\to\infty} \{\frac{1}{N} \ell n Q(NPT)\}$$

$$= -\lim_{N\to\infty} \{\ell n f_1 + \frac{1}{N} \ell n[1 + (\frac{f_2}{f_1})^N]\} \tag{2.11}$$

$$= -\lim_{N\to\infty} \{\ell n f_2 + \frac{1}{N} \ell n[1 + (\frac{f_1}{f_2})^N]\} \quad ,$$

or

$$\beta g(P,T) = \begin{cases} -\ell n f_1 \text{ for } f_1 > f_2 \\ -\ell n f_2 \text{ for } f_2 > f_1 \end{cases} \tag{2.12}$$

$$\beta g = M_{in}\{-\ell n f_1, -\ell n f_2\} \quad . \tag{2.13}$$

In this case we have two free energy surfaces. If the two surfaces cross over a certain curve, C, in the T-P plane a jump occurs from one surface to the other. The curve C is a locus of discontinuities (not singularities) along which we have a first order phase transition.

2. Instabilities and Singularities

A more common way in which transitions appear is illustrated by the following artificial example due to P. W. Kasteleyn [13].

Suppose that the canonical partition function for a specific model has the following exact form

$$Q(N,V,T) = \prod_{n=0}^{N-1} (x^2+1 - \frac{2n}{N} x) , \tag{2.14}$$

where

$$x = x(v,T) ,$$

is some intensive quantity.

Then,

$$\begin{aligned} \beta a(T,v) &= -\lim_{N\to\infty}\{\frac{1}{N} \ell nQ\} \\ &= -\lim_{N\to\infty}\{\frac{1}{N} \sum_{n=0}^{N-1} \ell n(x^2+1 - \frac{2n}{N} x) , \end{aligned} \tag{2.15}$$

or

$$\beta a = -\int_0^1 d\lambda \, \ell n(x^2+1 - 2\lambda x) = +\frac{1}{2x} \int_{x^2+1}^{(x-1)^2} \ell ny dy . \tag{2.16}$$

Carrying out the integration we have

$$\begin{aligned} \beta a(T,v) = \beta a(x) &= \frac{1}{2x} [y\ell ny-y]_{x^2+1}^{(x-1)^2} \\ &= \frac{1}{2x} \{(x-1)^2 \ell n(x-1)^2 - (x^2+1)\ell n(x^2+1)+2x\} . \end{aligned} \tag{2.17}$$

This function has a singularity at x=1, i.e. along the curve

$$x(T,v) = 1 ,$$

in the T-v plane. Let us examine derivatives of (βa) with respect to v to find the order of this transition.

$$\frac{\partial(\beta a)}{\partial v} = \frac{d(\beta a)}{dx}\frac{\partial x}{\partial v} \tag{2.18a}$$

$$\frac{\partial^2(\beta a)}{\partial v^2} = \frac{d^2(\beta a)}{dx^2}\left(\frac{\partial x}{\partial v}\right)^2 + \frac{d(\beta a)}{dx}\frac{\partial^2 x}{\partial v^2}\ , \tag{2.18b}$$

etc. Look at the derivatives with respect to x;one finds

$$\frac{d(\beta a)}{dx} = -\frac{\beta a}{x} + 2(x-1)\{\ell n(x-1)^2+1\} - 2x\{\ell n(x^2+1)+1\} - 2 \tag{2.19a}$$

$$\frac{d^2(\beta a)}{dx^2} = -\frac{2}{x}\frac{d(\beta a)}{dx} + 2\{\ell n(x-1)^2+1\} + 4$$

$$-2\{\ell n(x^2+1)+1\} - \frac{4x^2}{x^2+1}\ . \tag{2.19b}$$

Thus

$$\lim_{x\to 1} \beta a = -\{\ell n2+1\} \tag{2.20a}$$

$$\lim_{x\to 1} \frac{d(\beta a)}{dx} = -\ell n2 - 3\ , \tag{2.20b}$$

while

$$\frac{d^2(\beta a)}{dx^2} \underset{x\to 1+}{\sim} 4\ell n(x-1)+8\ . \tag{2.21}$$

The second derivatives of βa approach ∞ as $x\to 1$ from below and above, as shown in Figure 5.

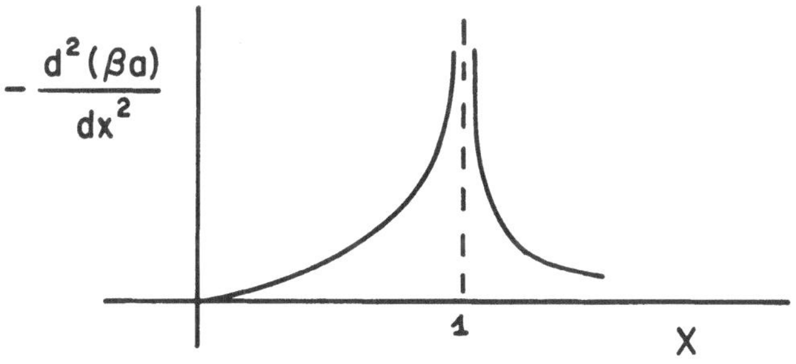

FIGURE 5. Divergence of $\frac{-d^2(\beta a)}{dx^2}$ for a simple model

This is the type of transition visualized in the Tisza Theory. In this case βa for $x>1$ has no analytical continuation beyond the point $x=1$ and the model does not describe any meta-stable states.

If, as we indicated, $x=x(v,T)$, this calculation would yield a locus of points $x(T,v) = 1$ along which $d^2(\beta a)/dx^2$ is infinite. As shown by Weeler and Griffiths [7] this would violate the conditions of thermodynamic stability. In this sense the example is too artificial.

C. Yang and Lee Theory

Yang and Lee [14] have shown how the possible occurrence of phase transitions can be related to the behavior of the zeros of the grand partition function. This theory has played such a central role in many papers on phase transitions that it should at least be mentioned in outline (an excellent account has been given by A. Münster [15]).

Let

$$\Xi(\lambda,T,V) = \sum_{N=0}^{B(V)} Q(N,V,T)\,\lambda^N \quad , \tag{2.22}$$

where $B(V)$ = maximum number of particles that can be packed into the volume V. Since $Q(N,V,T) > 0$ this polynomial will have no zeros on the positive real λ axis. However if we let (2.22) define Ξ for the entire complex λ plane we can introduce the following concept.

Def: A point λ_0 in the complex λ plane is called a limit point of zeros of Ξ if for every neighborhood, N, of λ_0 and every number Ω, there exists a volume $V>\Omega$ and a λ in N such that $\Xi(\lambda,V) = 0$.

Now the set of all limit points of the zeros of Ξ is thought to form a connected set of arcs in the complex λ plane. This has been true for every model for which one can solve for the zeros of Ξ [16,17,18].

Yang and Lee were able to show that a phase transition can occur only at those values of real λ where the set of limit points meets the real positive λ axis.

Pictorially, for some finite value of V we consider the B(V) zeros of Ξ

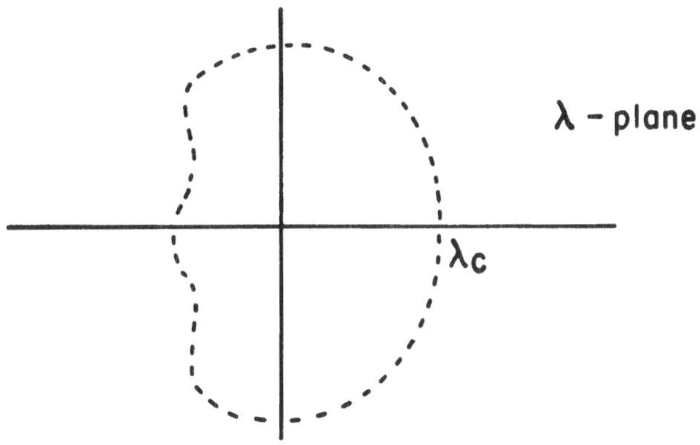

FIGURE 6. Zeros in the complex λ-plane of the grand partition function

As V→∞, B(V)→∞ and the number of zeros →∞. As the zeros get dense they may crowd in toward the real λ axis and this would give rise to a phase transition.

D. Some Recent Advances in the Statistical Mechanics of Phase Transitions

At this point it is appropriate to choose some special topic to discuss in detail. The list of possibilities is long and I do not want to spend the time making an exhaustive survey of recent advances. However there are a few major areas which should be mentioned for the sake of completeness. A list of some of the advances over the past ten years together with a few comments will also place our detailed topic in proper perspective.

1. Computer Calculations

Although computer calculations deal by necessity with systems that are finite in some sense, they have provided interesting information about phase transitions. There are three major methods in use:

(a) The Molecular Dynamics Method [19]

(b) The Monte Carlo Method [20]

(c) The Transfer Matrix Method [21]

In molecular dynamics calculations the equations of motion for N-particle systems (N usually less than 1,000) are integrated (subject to computer round-off error) and time averages of various quantities are computed over a calculated phase-space trajectory. One can now obtain numerical values for the equation of state outside of a phase transition region to an accuracy of 0.1%.

Using this method Alder discovered direct evidence for a first-order phase transition in a rigid sphere system (a transition predicted earlier by Kirkwood's approximate calculations). He later also established the evidence for a corresponding transition for a two dimensional rigid disk system [19].

In the Monte Carlo method configurations are generated by some probabilistic process (usually a Markov chain) with a distribution appropriate to the ensemble being considered. This method was used by W. W. Wood [20] to explore the hard sphere transition shortly after it was discovered by Alder and to establish the behavior of the rigid disk system concurrently with Alder.

It has also been used extensively to explore numerically the melting transition for a system of Lennard-Jones molecules.

We will discuss later in some detail the hard-sphere and hard-disk transitions and touch upon the question of how one extrapolates machine calculations on finite systems to the thermodynamic limit [21].

The transfer matrix technique is very different in character from the other numerical techniques. It proceeds not from basic principles but instead from the "transfer" matrix analysis of lattice systems [22]. The partition function of a lattice system, e.g. a two-dimensional (2-D) system of M rows and N columns, can be written as the trace of the Nth power of a transfer matrix

$$Q_{NM} = \mathrm{Trace}(\underset{\sim}{L}^N) \quad . \tag{2.23}$$

When the system becomes infinite in one direction ($N \to \infty$) Q_{NM} is given in terms of the largest eigenvalue λ_1 of $\underset{\sim}{L}$

$$Q_{NM} = \lambda_1^N \quad . \tag{2.24}$$

The computer calculation then involves the calculation of the largest eigenvalue of the large matrix $\underset{\sim}{L}$. Clever techniques using the symmetry of the problem have been devised to extend the range of values of M that can be handled.

The most interesting application thus far has been for systems of hard particles on a lattice. If such a particle is located on lattice site i then in the various models the 1st, 2nd, and 3rd etc. nearest neighbor sites of i are excluded from being occupied. These systems seem to display some type of transition which is 2nd or higher order when only 1st nearest neighbors are excluded but probably 1st order by the time 3rd neighbors are excluded (for more details see ref. 23-24).

2. Systems with Very Long Range Potentials

The belief has long persisted that the essential features of any theory of phase transitions is contained in the van der Waals and Weiss molecular field theories. Aside from detailed behavior or quantitative results for such things as the critical temperature there are physical arguments for believing that these theories contain an explanation for the existence of phase transitions.

Moreover there have been physical reasons why one should regard the van der Waals theory as representing the limiting behavior for a long-range attractive potential. That this is the case was made explicitly clear by a model introduced by Kac, Uhlenbeck and Hemmer [25]. They consider a pair potential of the form

$$Q(T) = q(R) + \gamma^{d}\psi(\gamma R)\,; \quad d = \text{dimension of space}\,, \tag{2.25}$$

where q(R) = short range force (usually taken to be a hard sphere potential) and where in one dimension $\psi(\gamma R)$ was written explicitly as

$$\psi(\gamma R) = \alpha e^{-\gamma R} \; . \tag{2.26}$$

The important feature of this work was that the authors were able to obtain an exact solution in one dimension by using the limit process $\gamma \to 0$.

The ideas introduced by Kac, Uhlenbeck and Hemmer have been treated in a general manner by Lebowitz and Penrose [26]. By considering the van der Waals limit ($\gamma \to 0$) for a certain class of potentials they found the general result in the thermodynamic limit

$$\lim_{\gamma \to 0} a(T,v,\gamma) = CE\{a^{o}(T,v) + \tfrac{1}{2}\alpha\rho^{2}\}\,, \tag{2.27}$$

where CE{f} denotes the maximum convex envelope of f and $\alpha = \int\psi(x)dx$.

3. Critical Phenomena

In early statistical mechanical studies of the critical point one will find most of the effort devoted to the calculation of such critical constants as $z_c = (p_c v_c / RT_c)$. More recently, the emphasis has been on the way thermodynamic quantities behave in the neighborhood of the critical point. This has led to the theory of the critical indices and the "scaling laws." One can say that this has been a truly active field and the work in it has produced many excellent reviews which should be consulted [G2, 27].

The concern with the critical region has also led workers into an investigation of the locus of plait points in a multicomponent system. As we know from the thermodynamic stability arguments, the infinities at the pure component critical points must become finite discontinuities along the plait locus. The behavior of properties near the plait points is a topic of current research.

4. Exactly Solvable Models

Simple models, for which exact solutions can be found, may tend to be abstract and rather remote as a model for a real physical system (witness the Ising Model for ferromagnetism) but they always invoke considerable interest. This is particularly true in the study of phase transitions, where the analytical properties of the partition functions are of extreme importance to our understanding. The insight we gain through such models can be obtained in no other way.

There have been several noteworthy advances in proposing special models in recent years. Just to mention two we have

(a) The ice model and related problems; the Rys F model of an antiferroelectric and the Slater KDP model of a ferroelectric. The common feature of these models is their reference to hydrogen-bonded crystals. E. Lieb [28] has been able to obtain exact solutions for the two-dimensional versions of these models.

(b) A model of two-dimensional polymer crystalization. This was solved exactly by Zwanzig and Lauritzen [29].

5. Rigorous Results of a General Nature

In the last decade the theory of Phase Transitions has been characterized more and more by careful and rigorous mathematical analysis of a general sort. Particularly pertinent are two fairly recent developments (see for other details the lectures by Professor Lebowitz and also ref. G.8).

(a) Some general proofs of the existence of first order phase transitions. The results seem to be limited to certain classes of lattice systems. They show, without recourse to special models, that first order phase transitions(constant P over a finite range in v, v_1, v_2) occur under well defined conditions [30].

(b) General proofs concerning the absence of long range order in one and two dimensions. Under well defined conditions a number of authors have been able to prove that in one and two dimensions long-range order cannot exist in the thermodynamic limit. This is not true of course, of the Ising model, but is true of the Heisenberg model, a two dimensional crystal and in several other cases [31]. We must point out that the absence of long-range order is not sufficient to exclude phase transitions in two-dimensional systems [40].

CHAPTER III. ABSENCE OF LONG-RANGE ORDER IN ONE-DIMENSION

A. Introduction

The one dimensional case points up the relation of two different questions.

1. Can long-range order be established as a state of thermodynamic equilibrium? In terms of an Ising-spin model this means an infinitely long sequence of + spins

+++++ .... +++

(or minus spins)

2. Can one achieve phase separation? In this case we consider two different states (e.g. cluster of (+) spins and clusters of (-) spins) and ask if in the limit of a very large system the free energy will be a minimum with the maximum separation of the phases (i.e. minimum surface).

B. Absence of Phase Separation

An argument can be given against the existence of any type of macroscopic phase in one dimension. Landau and Lifshitz present a thermodynamic version, but for completeness consider the following detailed binary lattice situation.

Consider a linear array of N equivalent sites. Each site may be occupied by a type A or a type B atom. We focus attention on the domains consisting of all A or B atoms and consider the case where there are (n+1) such domains.

e.g.

AAA BB A B A BB AAA B $\qquad\qquad$ $n+1 = 8$

(In fact let us assume that $\underline{n}$ is odd for the sake of convenience). Associate with each A-B boundary a "surface" energy 2w. Next define the following quantities

$$N_{AA} = \text{no. of A-A pairs}$$
$$N_{AB} = \text{no. of A-B pairs} \tag{3.1}$$
$$N_{BB} = \text{no. of B-B pairs}$$

with the usual consistency conditions,

$$2N_{AA} + N_{AB} = 2N_A$$
$$2N_{BB} + N_{AB} = 2N_B \quad . \tag{3.2}$$

If we only considered nearest neighbor interactions the total energy would be

$$E = N_{AA}G_{AA} + N_{AB}G_{AB} + N_{BB}G_{BB}$$
$$= (N_A - \tfrac{1}{2}N_{AB})G_{AA} + N_{AB}G_{AB} + (N_B - \tfrac{1}{2}N_{AB})G_{BB}$$
$$= N_AG_{AA} + N_BG_{BB} + N_{AB}(G_{AB} - \frac{G_{BB}+G_{AA}}{2}) \quad . \tag{3.3}$$

In this case

$$2w = G_{AB} - \tfrac{1}{2}(G_{AA}+G_{BB}) \quad . \tag{3.4}$$

Thus the "surface" energy, E_s, associated with this configuration is

$$E_s = N_{AB}(2w) \quad . \tag{3.5}$$

When n is odd

$$\text{No. of A domains} = \tfrac{1}{2}(N_{AB}+1) \tag{3.6}$$
$$\text{No. of B domains} = \tfrac{1}{2}(N_{AB}+1) \quad .$$

Thus

$$n = N_{AB} \tag{3.7}$$

and

$$E_s = 2nw \quad . \tag{3.8}$$

To compute the entropy we first compute the number of ways of arranging N_A A's into [(n+1)/2] domains. Now each domain must have at least one site occupied by an A. The question then is the number of ways of assigning the remaining X A's to Y domains, where

$$X = N_A - \frac{1}{2}(N_{AB}+1)$$
$$Y = \frac{1}{2}(N_{AB}+1) \quad . \tag{3.9}$$

We want to pick the Y objects (domains) X times with repetition, and then assign the A's. The count is

$$W_A = \binom{Y+X-1}{X} \cong \frac{N_A!}{(\frac{n}{2})!(N_A-\frac{n}{2})!} \quad . \tag{3.10}$$

(Neglect 1 compared to N_A,n). Similarly for the B's

$$W_B = \frac{N_B!}{(\frac{n}{2})!(N_B-\frac{n}{2})!} \quad . \tag{3.11}$$

Since the linear array can begin with an A or a B domain we have

$$W(n) = 2W_A W_B \quad , \tag{3.12}$$

and the entropy associated with a given value of n is

$$S = k \ln W(n) \quad . \tag{3.13}$$

Assuming that $1 << n << N_A$, N_B, we can expand the factorials using Stirling's approximation to obtain

$$S \sim -nk \ln \frac{n}{2} + \frac{nk}{2} \ln N_A N_B \quad . \tag{3.14}$$

The free energy is then

$$\Delta F = E_s - TS \sim n\{aw+kT\ln\frac{n}{2} - \frac{kT}{2} \ln N_A N_B\} \quad . \tag{3.15}$$

In the limit of a large system (N_A, $N_B \to \infty$) the free energy decreases as <u>n</u> increases which implies that phases cannot form, as the state with maximum free energy has maximum n, and hence a large number of domains of A and B. A macroscopic phase or domain would imply small n.

CHAPTER IV. LANDAU CRYSTALLINE STABILITY THEORY

A. Introduction

We now want to focus our attention on the crystalline state which we describe locally as follows.

Let $\rho(\vec{r})$ [the local density] be considered as a function of the spatial coordinates $\vec{r}$. On a macroscopic scale $\rho(\vec{r})$ will have the periodicity of the lattice in a crystalline solid. Although we shall be using thermodynamic arguments we consider $\rho(\vec{r})$ on this scale.

Let $a(T,p) = f(\vec{r})$ be the local Helmholtz free energy density. It is considered to be a function of ρ and thus implicitly a function of the spatial coordinates $\vec{r}$.

Landau [1] proposed the following thermodynamic argument to study the stability of the crystalline phase in 1, 2 and 3 dimensions.

Consider an isothermal fluctuation in which each volume element of material is displaced by an amount $\vec{s}(\vec{r})$. In other words we consider the strain in which

$$\vec{r} \rightarrow \vec{r} + \vec{s}(\vec{r}) = \vec{r}^{\,+} \quad . \tag{4.1}$$

Although the elements of the displacement vector are considered to arise through fluctuations we regard this as an infinitesimal strain and use the results of elasticity theory. The variations $d\vec{r}^{\,+}$ are related

$$\begin{aligned} d\vec{r}^{\,+} &= d\vec{r} + \frac{\partial \vec{s}}{\partial \vec{r}}\, d\vec{r} \\ &= d\vec{r}\,(\underset{\sim}{I}+\underset{\sim}{T}) \quad , \end{aligned} \tag{4.2}$$

by means of the strain tensor T

$$\underset{\sim}{T} = \frac{\partial \vec{s}}{\partial \vec{r}} \quad . \tag{4.3}$$

Moreover from elasticity theory we know that the free energy density depends upon the elements of the strain tensor

$$a^{+} = a + \sum_{iklm}\sum\sum\sum \lambda_{iklm} T_{ik} T_{lm} + \ldots \\ = A + \underset{\sim}{T}:\underset{\approx}{M}:\underset{\sim}{T} + \ldots \quad . \tag{4.4}$$

The stability criterion is obtained by examining the mean square fluctuation, σ^2, in the displacement $\vec{s}$

$$\sigma^2 = \langle \vec{s}\cdot\vec{s} \rangle \quad , \tag{4.5}$$

where the average has to be defined more precisely. If σ^2 remains $<\infty$ for an infinite system then the structure is assumed stable, but if $\sigma^2 \to \infty$ as $N,V\to\infty$ then the crystal is regarded as unstable. The first case happens in three dimensions whereas the second occurs for one and two dimensional systems (see also [34] and below for more details).

B. Review of Fluctuation Theory

Consider an isolated system (corresponding to a microcanonical ensemble). A fluctuation in the system will be accompanied by a change in entropy ΔS_t which is negative. The thermodynamic theory of fluctuations says that the probability density for observing this fluctuation is given by

$$W \propto e^{\Delta S_t/k} \quad . \tag{4.6}$$

To examine fluctuations in a closed isothermal system, we consider such a system, s, surrounded by a bath B. Now let a fluctuation occur in s. The total change in entropy of s is

$$\Delta S = \Delta_e S + \Delta_i S \; , \tag{4.7}$$

where $\Delta_e S = \frac{đq}{T}$ = flow of entropy through the boundaries of s and

$\Delta_i S > 0$ is the entropy production.

For the bath B, the change is reversible and

$$\Delta S_B = \Delta_e S_B = \frac{đq}{T} \quad . \tag{4.8}$$

For the total isolated system then

$$\Delta S_t = \Delta S_s + \Delta S_B = \Delta_i S = \Delta S_s - \frac{đq}{T} = \Delta S_s - \frac{1}{T}(\Delta E + p\Delta V) \quad . \tag{4.9}$$

Therefore

$$\frac{\Delta S_s}{k} = -\frac{1}{k_B T}(\Delta E - T\Delta \quad + p\Delta V) \quad , \tag{4.10}$$

and when T and V are constant one finds

$$w \alpha e^{-\Delta A/k_B T} \tag{4.11}$$

$$A = E - TS_s \ ,$$

and in terms of the free energy density

$$\Delta A = \int (a^+ - a) d\vec{r} \quad . \tag{4.12}$$

C. Stability Analysis

The instability which develops in one and two dimensional crystals is in the long wavelength displacements (locally crystal structure is maintained.) To see this clearly we expand $\vec{s}$ as a Fourier series.

Let V be in the form of a cube with sides L and consider the system under periodic boundary conditions. Then*

$$\vec{s} = \sum_{\substack{\vec{k} \\ \neq 0}} \vec{u}_k e^{i\vec{k}\cdot\vec{r}} \quad ; \quad \vec{k} = \frac{2\pi}{L}\vec{n} \quad . \tag{4.13}$$

Since $\vec{s}$ is real we know that

$$\vec{u}_{\vec{k}}^{\,*} = \vec{u}_{-\vec{k}} \quad . \tag{4.14}$$

The strain tensor $\underset{\sim}{T}$ has the following expansion

$$\underset{\sim}{T} = \frac{\partial\vec{s}}{\partial\vec{r}} = i \sum_{\vec{k}\neq 0} \vec{k}\, \vec{u}_{\vec{k}}\, e^{i\vec{k}\cdot\vec{r}} \quad , \tag{4.15}$$

and the free energy density becomes

*Note k ≠ 0 because we omit a uniform displacement which amounts to a translation.

$$a^{+}-a = -\sum_{\vec{k}}\sum_{\vec{k}'} \vec{k}\,\vec{u}_{\vec{k}} : \underset{\approx}{M} : \vec{k}'\,\vec{u}_{\vec{k}'}\, e^{i(\vec{k}+\vec{k}')\cdot\vec{r}} . \qquad (4.16)$$

The total free energy change is then

$$\Delta A_t = -\sum_{\vec{k}}\sum_{\vec{k}'} \vec{k}\,\vec{u}_{\vec{k}} : \underset{\approx}{M} : \vec{k}'\,\vec{u}_{\vec{k}'} \int e^{i(\vec{k}+\vec{k}')\cdot\vec{r}}\, d\vec{r} . \qquad (4.17)$$

However we can make use of the formal identity

$$\delta(\vec{k}+\vec{k}') = \delta_{\vec{k},-\vec{k}} = \frac{1}{V}\int e^{i(\vec{k}+\vec{k}')\cdot\vec{r}}\, d\vec{r} , \qquad (4.18)$$

to write

$$\Delta A_t = +V \sum_{\vec{k}} \vec{k}\,\vec{u}_{\vec{k}} : \underset{\approx}{M} : \vec{u}^{*}_{\vec{k}}\,\vec{k} \qquad (4.19)$$

The probability of observing a fluctuation now takes the form

$$w = K\, e^{\frac{-V}{KT}\sum_{k} \vec{u}_k\cdot \underset{\sim}{B}\cdot \vec{u}^{*}_k} \qquad (4.20)$$

$$\underset{\sim}{B} = \underset{\sim}{B}(\vec{k}) = \vec{k}\cdot\underset{\approx}{M}\cdot\vec{k} = k^2\, \vec{e}\cdot\underset{\approx}{M}\cdot\vec{e} = k^2\, \underset{\sim}{\hat{B}} , \qquad (4.21)$$

where $\vec{e}$ is the unit vector in the $\vec{k}$ direction. The fluctuations in the Fourier components are independent of each other (i.e. $\underline{w}$ factors into a product of probability densities).

Thus

$$\langle \vec{s}\cdot\vec{s}\rangle = \langle \sum_{\vec{k}}\sum_{\vec{k}'} \vec{u}_{\vec{k}} \cdot \vec{u}^{*}_{\vec{k}'}\, e^{i(\vec{k}-\vec{k}')\cdot\vec{r}} \rangle$$

$$= \sum_{\vec{k}}\sum_{\vec{k}'} \langle \vec{u}_{\vec{k}}\cdot\vec{u}_{\vec{k}'} \rangle\, e^{i(\vec{k}-\vec{k}')\cdot\vec{r}} . \qquad (4.22)$$

Moreover

$$\langle \vec{u}_{\vec{k}}\cdot\vec{u}^{*}_{\vec{k}'}\rangle = \langle \vec{u}_{\vec{k}} \rangle\, \langle \vec{u}^{*}_{\vec{k}'}\rangle \quad \text{for } \vec{k} \neq \pm\vec{k}' , \qquad (4.23)$$

and

$$\langle \vec{u}_{\vec{k}}\rangle = 0 \qquad (4.24)$$

since $\langle\vec{s}\rangle = 0$ for an equilibrium state and all possible fluctuations. This means that

$$\langle s^2 \rangle = \sum_{\substack{\vec{k} \\ \neq 0}} \langle |\vec{u}_k|^2 \rangle . \tag{4.25}$$

We now examine the averages of the individual Fourier components:

$$\langle \vec{u}_k \cdot \vec{u}_k^* \rangle = c \int du_k \, u_k \cdot u_k \, e^{-\beta V k^2 \vec{u}_k \cdot \hat{\underset{\sim}{B}} \cdot \vec{u}_{\vec{k}}^*}$$

$$= \frac{1}{\beta V k^2} \frac{1}{|\hat{\underset{\sim}{B}}|} , \tag{4.26}$$

where $\hat{B}$ depends upon the unit vector in the $\vec{k}$ direction. Returning to Eq. (4.25) we find that

$$\langle s^2 \rangle = \frac{1}{\beta} \frac{1}{V} \sum_{\substack{k \\ \neq 0}} \frac{1}{k^2 |\hat{\underset{\sim}{B}}|} . \tag{4.27}$$

As V, L → ∞ the sum over $\vec{k}$ becomes an integral and we find

$$\langle s^2 \rangle = \frac{1}{\beta} \int d\vec{k} \, \frac{1}{k^2 |\hat{\underset{\sim}{B}}|} = \int \frac{k^{d-1} dk}{k^2} \int d\vec{e} \, \frac{1}{|\hat{B}|} , \tag{4.28}$$

where d = dimension of $\vec{k}$ (i.e. of the system). Thus we see that for one and two dimensions

$$\langle s^2 \rangle \to \infty , \tag{4.29}$$

in the thermodynamic limit (see also for other details [34] and [15] § 5.12 p. 387.)

CHAPTER V. LONG-RANGE CRYSTALLINE ORDER

A. Introduction

In approaching a problem in the solid state one is immediately faced with the necessity of defining a criterion for the presence of ordering. The starting point for such considerations is the singlet density or one particle distribution function. Confining our considerations to classical statistical mechanics we start by defining

$$\nu(\vec{r}) = \sum_{i=1}^{N} \delta(\vec{r}-\vec{R}_i) \quad , \tag{5.1}$$

which is a microscopic singlet density in which $\vec{r}$ is an arbitrary point in space and $\vec{R}_i$ is the position vector of particle i. From (5.1) we define

$$\rho(\vec{r}) = \langle\nu(\vec{r})\rangle \quad , \tag{5.2}$$

where to be specific we shall take the average in the ordinary canonical ensemble.

One would like to consider a formulation in which $\rho(\vec{r})$ for a perfect crystal would be a periodic function possessing the long-range order characteristic of the space lattice of the crystal. However, if one introduces only the usual potential energy

$$\Phi_1(\vec{R}_1,\ldots,\vec{R}_N) = \sum_{i<j} \phi(R_{ij}) \quad , \tag{5.3}$$

which is translationally and rotationally invariant, then $\rho(\vec{r})$ will have the same property and be a uniform constant. To overcome this objection let us first consider N point particles contained in a cubic box V with periodic boundary conditions at a temperature T and with a total potential Φ being the sum of two parts

$$\Phi(\vec{R}_1,\ldots,\vec{R}_N) = \Phi_1(\vec{R}_1,\ldots,\vec{R}_N) + \Phi_2(\vec{R}_1,\ldots,\vec{R}_N) \, , \tag{5.4}$$

where Φ_1 is the usual intermolecular potential function defined in

(5.3) and Φ_2 is an external potential preserving any desired crystal configuration

$$\Phi_2 = \sum_{i=1}^{N} \varepsilon_N \, \phi_2(\vec{R}_i) , \tag{5.5}$$

Φ_2 is taken to be a sum of singlet potential functions. Note, first of all, that only the six orientational and translational degrees of freedom of the entire crystal need to be frozen. Since each degree of freedom will have an average energy of the order of $k_B T$ the magnitude of each term in Φ_2 need only be of the order $6k_B T/N$. As N and V become arbitrarily large each term in Φ_2 can be made smaller and smaller. Thus

$$\varepsilon_N \to 0 \quad \text{as } N \to \infty, \text{ and } ,$$

Φ_2 presumably would have periodic minima at the equilibrium crystal sites.

For a finite system we can expand $\rho(\vec{r})$ in a three dimensional Fourier Series.

$$\rho(\vec{r}) = \sum_{\vec{k}} \rho_{\vec{k}} \, e^{i\vec{k}\cdot\vec{r}} , \tag{5.6}$$

with

$$\vec{k} = \frac{2\pi\vec{n}}{L} \quad ; \quad L = V^{1/d} \quad ; \quad \vec{n} = \text{integer vector} .$$

It is also important to remember that in the thermodynamic limit

$$\frac{1}{V} \sum_{k \neq 0} \xrightarrow[V \to \infty]{} \int_{-\infty}^{\infty} d\vec{k} . \tag{5.7}$$

The following relations also exist :

$$\begin{aligned} \rho_{\vec{k}} &= \frac{1}{V} \int \rho(\vec{r}) e^{-i\vec{k}\cdot\vec{r}} d\vec{r} \\ &= \frac{1}{V} \int \langle \nu(\vec{r}) \rangle e^{-i\vec{k}\cdot\vec{r}} d\vec{r} \\ &= \langle \nu_{\vec{k}} \rangle = \frac{1}{V} \sum_{i=1}^{N} \langle e^{-i\vec{k}\cdot\vec{R}_i} \rangle , \end{aligned} \tag{5.8}$$

where

$$\nu_{\vec{k}} = \frac{1}{V} \sum_{i=1}^{N} e^{-i\vec{k}\cdot\vec{R}_i} \tag{5.9}$$

B. Lattice Vectors and Reciprocal Lattice Vectors

In a perfect infinite crystal we have an additional symmetry requirement on $\rho(\vec{r})$ due to the translational symmetry of the lattice. Define

$$\vec{a}_i \quad i=1,\ldots,d \quad ,$$

as a set of basic lattice vectors. Then

$$\vec{T}_{\vec{n}} = n_1\vec{a}_1 + \ldots + n_d\vec{a}_d \ ,$$

with $n_1,\ldots,n_d = 0,1,2,\ldots$.will be a translation vector for which

$$\rho(\vec{r}+\vec{T}) = \rho(\vec{r}) \quad . \tag{5.10}$$

More specifically ρ must be periodic in $\vec{r}$ for all $\vec{a}_i$

Next define a set of reciprocal lattice vectors

$$\vec{b}_i \ ; \ i=1\ldots d \quad ,$$

such that

$$\vec{b}_i\cdot\vec{a}_j = 2\pi\ \delta_{ij} \quad . \tag{5.11}$$

Then if each Fourier component of $\rho(\vec{r})$ is to have the symmetry of the lattice we require that

$$\vec{k} = \vec{K} = m_1\vec{b}_1+\ldots+m_d\vec{b}_d \ ; \ \vec{K}\cdot\vec{T} = 2\pi x(\text{integer}) \tag{5.12}$$

[i.e. $\vec{k}\cdot\vec{a}_i = 2\pi m_i$ for all i] ,

where $m_1,\ldots m_d$ are integers. The vector defined in (5.12) is called a reciprocal lattice vector.

C. Long-Range Crystallinity

We formulate now the following criterion for a crystal to exhibit long-range order:

In the Fourier decomposition of the singlet distribution function:

$$\rho(\vec{r}) = \sum_{\vec{k}} \rho_{\vec{k}} e^{i\vec{k}\cdot\vec{r}} ,$$

the Fourier coefficients, $\rho_{\vec{k}}$, have the following property:

$\rho_{\vec{k}} = 0$ if $\vec{k} \neq \vec{K}$ a reciprocal lattice vector.

$\rho_{\vec{k}} \neq 0$ for at least one non-zero reciprocal lattice vector $\vec{k} = \vec{K}$.

To demonstrate the absence of long-range crystalline order in 1 and 2 dimensions it is sufficient to show that

$$\rho_{\vec{K}} \to 0 \quad \text{as } N,V \to \infty \quad , \tag{5.13}$$

for all non-zero reciprocal lattice vectors , $\vec{k}$.

We look for an inequality

$$A \geq B \quad ,$$

in which

(a) $\lim_{\substack{N,V \to \infty \\ \frac{N}{V} = \text{const.}}} A = A^{*} < \infty$

and (b) $B = C\, K^2 \rho^2_{\vec{k}} \{\frac{1}{V} \sum_{\vec{k}\neq 0} \frac{1}{k^2}\}$,

where C is a constant and $\vec{K}$ is a reciprocal lattice vector. Taking the limit $N,V \to \infty$ gives

$$\infty > A^{*} \geq C\, K^2 \rho^2_{\vec{K}} \int_{-\infty}^{\infty} \frac{d\vec{k}}{k^2} .$$

The integral diverges in one and two dimensions which implies that

$$|\rho_{\vec{K}}| \to 0 \quad \text{for } \vec{K} \neq 0 \ .$$

The establishment of the appropriate inequality requires some judicious and obscure insight, but some common features of these types of proof are demonstrated.

D. Basic Inequalities

The method used to show the existence or not of a phase transition makes use of certain basic inequalities called the Gibbs-Bogoliubov inequality and the Bogoliubov inequality. Before proceeding we will review these inequalities.

The Gibbs-Bogoliubov inequality has been so widely used to obtain bounds on the exact free energy that we think it is worthwhile to review two of the simpler methods of derivation.

A derivation given by Falk [32] makes use of the convexity of the function

$$f(x) = e^{-\beta x} \quad , \tag{5.14}$$

First expand $f(x)$ in the neighborhood of a point $\bar{x}$

$$f(x) = f(\bar{x}) + f^{(1)}(\bar{x})(x-\bar{x}) + \frac{1}{2} f^{(2)}(\bar{x})(x-\bar{x})^2 \tag{5.15}$$

Note that $f^{(2)}(\bar{x}) = \beta^2 e^{-\beta\bar{x}} > 0$. Thus

$$f(x) \geq f(\bar{x}) + f^{(1)}(\bar{x})(x-\bar{x}) \ . \tag{5.16}$$

We next consider the mean of the function $f(x)$ for an arbitrary distribution

$$\langle f(x)\rangle \geq f(\bar{x}) + f^{(1)}(\bar{x})(\langle x\rangle-\bar{x}) \ . \tag{5.17}$$

If we choose $\bar{x} = \langle x\rangle$ then

$$\langle f(x)\rangle \gtrsim f(\langle x\rangle)\,, \tag{5.18}$$

the basic inequality we want to use.

In particular we let $f = e^{-\beta H}$ with $H = H_0+H_1$. Define $\langle\ldots\rangle_0$ as the trace or the sum over all unperturbed eigenstates. Then using (5.18) we obtain (note that this proof is in the quantum case)

$$\langle e^{-\beta H}\rangle_0 \geq e^{-\beta\langle H_0+H_1\rangle_0} \tag{5.19}$$

$$e^{-\beta A} \geq e^{-\beta A_0-\beta\langle H_1\rangle_0}\quad , \tag{5.20}$$

where $e^{-\beta A} = \langle e^{-\beta H}\rangle$. Note $\langle e^{-\beta H_0}\rangle_0 = e^{-\beta\langle H_0\rangle_0}$ since we are summing over unperturbed eigenstates. Then $A\leq A_0 + \langle H_1\rangle_0$.

Next, we shall present a derivation due to Isihara [33]. Let $x = \{p^N,q^N\}$ be a point in phase space. Then consider two phase-space distribution functions $f(x)$ and $g(x)$ both of which are positive and properly normalized

$$\int f(x)dx = \int g(x)dx = 1\quad . \tag{5.21}$$

The basic inequality we want to work with can then be expressed in the following form

$$\int f(x)\ln f(x)dx \geq \int f(x)\ln g(x)dx\ , \tag{5.22}$$

or $$\langle \ln f\rangle_f \geq \langle \ln g\rangle_f\quad ,$$

which may look strange since g and f are arbitrary distribution functions.

To demonstrate that this inequality holds we start with the elementary one

$$y\ln y-y \gtrsim -1 \qquad \text{for } y > 0\ .$$

Next consider

$$J = \int f(x)\ln f(x)dx - \int f(x)\ln g(x)dx \tag{5.23}$$

$$= \int g\{\frac{f}{g}\ln f - \frac{f}{g}\ln g\}dx \tag{5.24}$$

$$= \int g\{\frac{f}{g}\ln\frac{f}{g} - \frac{f}{g} + 1\}dx \geq 0 \quad .$$

This proves equation (5.22).

To convert this inequality into our working form we consider the hamiltonian

$$H = H_o + H_1 \quad , \tag{5.25}$$

and the corresponding free energy, A, which we write as

$$A = A_o + A_1 \quad , \tag{5.26}$$

where A_o is the free energy for the hamiltonian H_o. Let

$$f = e^{\beta(A_o - H_o)}$$
$$g = e^{\beta(A-H)} \quad . \tag{5.27}$$

Then

$$\langle \ln f \rangle_f \geq \langle \ln g \rangle_f \quad ,$$

or

$$\beta(A_o - \langle H_o \rangle_o) \geq \beta(A - \langle H \rangle_o)$$

$$\text{or } \langle H - H_o \rangle_o \geq A_1 \quad . \tag{5.28}$$

A_1 can also be bounded from below by inverting the roles of f and g and then we obtain

$$\langle A - H \rangle \geq \langle A_o - H_o \rangle \quad , \tag{5.29}$$

or

$$A_1 \geq \langle H - H_o \rangle = \langle H_1 \rangle \quad . \tag{5.30}$$

Thus in summary

$$\langle H_1 \rangle_o \geq A_1 \geq \langle H_1 \rangle \quad . \tag{5.31}$$

E. Bogoliubov Inequality-Classical Version

Consider two phase-space functions $A(x)$ and $\vec{B}(x)$ where $\vec{B}$ is a vector function. The canonical averages of these functions and the averages of their products satisfy a Schwarz inequality.

$$<|A|^2> <|\vec{B}|^2> \geq |<A^*\vec{B}>|^2 \quad . \tag{5.32}$$

The inequality can be established as follows. Consider:

$$<|B - \frac{A}{<|A|^2>} <A^*\vec{B}>|^2> \geq 0 \quad . \tag{5.33}$$

Expanding the square one obtains

$$<\vec{B}^*\cdot\vec{B} - \frac{A\vec{B}^*}{<|A|^2>} \cdot <A^*\vec{B}> - \frac{A^*\vec{B}}{<|A|^2>} \cdot <A\vec{B}^*> + \frac{A^*A}{<|A|^2>^2} <A^*\vec{B}> \cdot <A\vec{B}^*>$$

$$= <|\vec{B}|^2> - \frac{|<A^*\vec{B}>|^2}{<|A|^2>} \geq 0 \quad , \tag{5.34}$$

which demonstrates the validity of the general Schwarz inequality. Practical use of Eq. (5.32) depends upon a judicious choice of A and $\vec{B}$ which in each case is by no means obvious. Let us consider the special choices introduced by N. D. Mermin [31].

Let

$$\vec{B} = -\frac{e^{\beta\Phi}}{\beta} \sum_{i=1}^{N} \frac{\partial}{\partial \vec{R}_i} [\phi_i e^{-\beta\Phi}] \quad , \tag{5.35}$$

where Φ is the potential energy and ϕ_i is a function of R_i which (together with Φ) obeys periodic boundary conditions. Also let

$$A = \sum_{i=1}^{N} \psi(\vec{R}_i) = \sum_i \psi_i \quad . \tag{5.36}$$

Then the inequality (5.32) can be put in the form

$$<|\sum_i \psi_i|^2> \geq \frac{|\sum_i <\phi_i \vec{\nabla}_i \psi_i^*>|^2}{\sum_{i,j=1}^{N} \beta<\phi_i \phi_j^* \vec{\nabla}_i \vec{\nabla}_j \Phi> + \sum_{i=1}^{N} <|\vec{\nabla}_i\phi_i|^2>} \geq 0 \quad . \tag{5.37}$$

Some of the physical motivation behind this approach is to introduce an externally applied periodic potential like

$$\phi_i = \sin \vec{k}\cdot\vec{R}_i \quad ,$$

and test the "crystalline" response to this potential. In particular one wants to look at the reciprocal lattice components of $\rho(\vec{r})$

$$\psi_i = e^{i\vec{K}.\vec{R}_i} \quad .$$

In the actual proof we shall take

$$\phi_i = \sin \vec{k}\cdot\vec{R}_i \ ; \qquad \vec{k} = \frac{2\pi}{L}\vec{n} \quad , \tag{5.38}$$

and

$$\psi_i = e^{i(\vec{k}+\vec{K})\cdot\vec{R}_i} \ ; \ \vec{K} = \text{reciprocal lattice vector} \ .$$

Both functions ϕ_i and ψ_i have the period of the cube of side L. In order to verify eq. (5.37), first consider quantity $<\vec{B}X>$

$$<\vec{B}X> = -\frac{1}{\beta\phi}\int d\vec{R}^N \sum_{i=1}^{N} X\frac{\partial}{\partial\vec{R}_i}[\phi_i\, e^{-\beta\Phi}]$$

$$= -\frac{1}{\beta\phi}\sum_i [X\ \phi_i e^{-\beta\Phi}]_o^L - \int d\vec{R}^N \phi_i e^{-\beta\Phi}\frac{\partial X}{\partial\vec{R}_i}$$

$$= \frac{1}{\beta} < \phi(\vec{R}_i)\frac{\partial X}{\partial\vec{R}_i}> \quad . \tag{5.39}$$

Applying this result to A^* and $\vec{B}^*$ we have

$$<\vec{B}A^*> = \frac{1}{\beta}\sum_{i=1}^{N} <\phi_i\ \frac{\partial\psi_i^*}{\partial\vec{R}_i}> \quad , \tag{5.40}$$

and

$$<\vec{B}\cdot\vec{B}^*> = \frac{1}{\beta}\sum_{i=1}^{N} <\phi_i\ \vec{\nabla}_i\cdot\vec{B}^N> \quad . \tag{5.41}$$

To obtain the reduction of $<\vec{B}\cdot\vec{B}^*>$ expand eq. (5.35) as

$$B^* = -\frac{1}{\beta}\sum_{j=1}^{N}\frac{\partial\phi_j^*}{\partial\vec{R}_j} + \sum_{j=1}^{N}\phi_j^*\ \frac{\partial\Phi}{\partial\vec{R}_j} \quad . \tag{5.42}$$

Then we compute

$$\vec{\nabla}_i \cdot \vec{B}^* = -\frac{1}{\beta} \nabla_i^2 \phi_i^* + (\vec{\nabla}_i \phi_i^*) \cdot (\vec{\nabla}_i \Phi) + \sum_{j=1}^{N} \phi_j \vec{\nabla}_i \cdot \vec{\nabla}_j \Phi ,$$

and from (5.41) form

$$\langle \vec{B} \cdot \vec{B}^* \rangle = \sum_{i=1}^{N} \frac{1}{\beta} \langle \phi_i \vec{\nabla}_i \phi_i^* \cdot \vec{\nabla}_i \Phi - \frac{\phi_i}{\beta} \vec{\nabla}_i^2 \phi_i^* \rangle$$

$$+ \frac{1}{\beta} \sum_i^N \sum_j \langle \phi_i \phi_j^* \vec{\nabla}_i \vec{\nabla}_j \Phi \rangle \quad . \tag{5.43}$$

The first term in (5.43) can be further simplified by another integration by parts. Since

$$\frac{\partial}{\partial \vec{R}_i} \cdot [e^{-\beta\Phi} \vec{\nabla}_i \phi_i^*] = e^{-\beta\Phi} \{ \vec{\nabla}_i^2 \phi_i^* - \beta (\vec{\nabla}_i \phi_i^*) \cdot \vec{\nabla}_i \Phi \} ,$$

we can write

$$\langle \beta \phi_i \vec{\nabla}_i \phi_i \cdot \vec{\nabla}_i \Phi - \phi_i \vec{\nabla}_i^2 \phi_i^* \rangle = -\frac{1}{\phi} \int d\vec{R}^N \phi_i \frac{\partial}{\partial \vec{R}_i} \cdot [e^{-\beta\Phi} \vec{\nabla}_i \phi_i^*] = \langle |\vec{\nabla}_i \phi_i|^2 \rangle \tag{5.44}$$

and

$$\langle \vec{B} \cdot \vec{B}^* \rangle = \frac{1}{\beta} \sum_i \sum_j \langle \phi_i \phi_j^* \vec{\nabla}_i \cdot \vec{\nabla}_j \Phi \rangle + \frac{1}{\beta^2} \sum_{i=1}^{N} \langle |\vec{\nabla}_i \phi_i|^2 \rangle . \tag{5.45}$$

Using (5.45) for $\langle \vec{B} \cdot \vec{B}^* \rangle$ and (5.40) for $\langle \vec{B} A^* \rangle$ in the inequality

$$\langle |A|^2 \rangle \geq \frac{|\langle A^* B \rangle|^2}{\langle \vec{B}^2 \rangle} \geq 0 \quad , \tag{5.46}$$

we obtain the result given in (5.37).

F. <u>Mermin Analysis</u>

We start with the following basic inequality

$$\langle |\sum_{i=1}^{N} \psi_i|^2 \rangle \geq \frac{|\sum_i \langle \phi_i \vec{\nabla}_i \psi_i^* \rangle|^2}{\langle \sum_i \sum_j \beta \phi_i \phi_j^* \vec{\nabla}_i \cdot \vec{\nabla}_j \Phi + \sum_i |\vec{\nabla}_i \phi_i|^2 \rangle} \geq 0 , \tag{5.47}$$

and set

$$\psi_i = e^{-i(\vec{k}+\vec{K})\cdot R_i}$$
(5.48)
$$\phi_i = \sin \vec{k}\cdot\vec{R}_i \quad .$$

(i) Analysis of $<|\Sigma\psi_i|^2>$

Letting

$$\vec{q} = \vec{k}+\vec{K} \tag{5.49}$$

we write

$$<|\Sigma\psi_i|^2> = <\sum_i \sum_j e^{-i\vec{q}\cdot\vec{R}_i-\vec{R}_j)}> \quad . \tag{5.50}$$

Referring to Equation (4.9) this can be expressed in terms of the $\nu_{\vec{k}}$, namely

$$<|\Sigma\psi_i|^2> = V^2 <\nu_{\vec{q}}\nu_{-\vec{q}}> \quad , \tag{5.51}$$

and we see that this term is related to a type of static structure factor. To see this consider the correlation function

$$C_N(\vec{r},\vec{r}') = <\nu(\vec{r})\nu(\vec{r}')> \quad , \tag{5.52}$$

and its Fourier transform

$$S_N(\vec{q}) = \frac{1}{V}\int d\vec{r}\int d\vec{r}'\, e^{-i\vec{q}\cdot(\vec{r}-\vec{r}'}C_N(\vec{r},r')$$

$$= \frac{1}{V}<\int d\vec{r}\int d\vec{r}'\, e^{-i\vec{q}\cdot(\vec{r}-\vec{r}')}\sum_{i,j}\delta(\vec{r}-\vec{R}_i)\delta(\vec{r}-\vec{R}_j)> \tag{5.53}$$

$$= \frac{1}{V}<\sum_i\sum_j e^{-i\vec{q}\cdot(\vec{R}_i-\vec{R}_j)}> \quad .$$

In the thermodynamic limit we define an

$$S(\vec{q}) = \lim_{\substack{N,V\to\infty \\ \frac{N}{V}=\rho}} S_N(\vec{q}) \quad , \tag{5.54}$$

and note that

$$<|\Sigma\phi_i|^2> = VS_N(\vec{q}) \quad . \tag{5.55}$$

(ii) <u>Analysis of $|\sum_i <\phi_i \vec{\nabla}\psi_i^*>|^2$</u>

Note first that

$$\phi_i \vec{\nabla}\psi_i^* = \frac{1}{2i}[e^{i\vec{k}\cdot\vec{R}} - e^{-i\vec{k}\cdot\vec{R}}]i(\vec{k}+\vec{K})e^{+i(\vec{k}+\vec{K})\cdot\vec{R}}$$

$$= \frac{1}{2}(\vec{k}+\vec{K})[e^{i(2\vec{k}+\vec{K})\cdot\vec{R}} - e^{i\vec{K}\cdot\vec{R}}] . \tag{5.56}$$

Then

$$\sum_i <\phi_i \vec{\nabla}_i \psi_i^*> = \frac{1}{2}(\vec{k}+\vec{K})[\rho_{2\vec{k}+\vec{K}} - \rho_{\vec{K}}]V$$

and

$$|\sum_i <\phi_i \vec{\nabla}_i \psi_i^*>|^2 = \frac{V^2}{4}(\vec{k}+\vec{K})^2 (\rho_{\vec{K}} - \rho_{2\vec{k}+\vec{K}})^2 . \tag{5.57}$$

Putting Eqs. (5.55) and (5.57) into the basic inequality (5.45) we have at this point

$$S_N(\vec{q}) \geq \frac{\frac{1}{4}(\vec{k}+\vec{K})^2(\rho_{\vec{K}}-\rho_{2\vec{k}+\vec{K}})^2}{\frac{1}{V}<\sum_i \sum_j \beta\phi_i\phi_j^* \vec{\nabla}_i\cdot\vec{\nabla}_j \Phi + \sum_i |\vec{\nabla}_i\phi_i|^2>} . \tag{5.58}$$

We also note that

$$\nabla_i \phi_i = \vec{k} \cos^2 \vec{k}\cdot\vec{R}_i \quad , \tag{5.59}$$

so that we can write

$$S_N(\vec{k}+\vec{K}) \geq \frac{1}{4} \frac{(\vec{k}+\vec{K})^2}{k^2} \frac{(\rho_{\vec{K}} - \rho_{2\vec{k}+\vec{K}})^2}{\frac{1}{V}<V_N + \beta\Gamma_N>} , \tag{5.60}$$

where

$$V_N = \sum_{i=1}^{N} \cos^2 \vec{k}\cdot\vec{R}_i \tag{5.61}$$

$$\Gamma_N = \sum_i \sum_j \frac{\sin \vec{k}\cdot\vec{R}_i}{k} \frac{\sin \vec{k}\cdot\vec{R}_i}{k} \vec{\nabla}_i\cdot\vec{\nabla}_j\Phi \quad .$$

(iii) Behavior of Denominator in Thermodynamic Limit

The question of the existence of $\frac{1}{V}\langle V_N+\beta\Gamma_N\rangle$ in the thermodynamic limit plays a role in the analysis, and this can be discussed in terms of the Gibbs-Bogoliubov inequality.

(1) $$\frac{1}{V}\langle V_N\rangle = \frac{1}{V}\langle\sum_i \cos^2\vec{k}\cdot\vec{R}_i\rangle \leq \frac{1}{V}\langle\sum_{i=1}^{N} 1\rangle$$

$$= \frac{N}{V} \rightarrow \rho \quad \text{as } N,V \rightarrow\infty \quad . \tag{5.62}$$

(2) $$\frac{\beta}{V}\langle\Gamma_N\rangle < \infty .$$

Consider a system whose intermolecular potential is given by

$$V_N = \Phi_N - \Gamma_N . \tag{5.63}$$

Then by the Gibbs-Bogoliubov inequality

$$\frac{1}{V}\langle\Gamma_N\rangle \leq \frac{1}{V}\{A_{\Phi_N} - A_{V_N}\}$$

$$= \rho\{\frac{A_\Phi}{N} - \frac{A_V}{N}\} \tag{5.64}$$

where A_Φ is the Helmholtz free energy computed from the potential Φ_N and A_V is the Helmholtz free energy computed from the potential $\Phi_N-\Gamma_N$.

If Φ and Γ obey sufficient conditions for the existence of the specific free energy then

$$\lim_{N,V\rightarrow\infty} \frac{1}{V}\langle\Gamma_N\rangle < \rho\{a_\Phi - a_V\} < \infty . \tag{5.65}$$

Thus $\frac{1}{V}\langle\Gamma_N\rangle$ will exist in the thermodynamic limit.

(iv) Γ_N as a sum of Pair Potentials

Let

$$\Phi = \sum_{i<j} \phi_{ij} = \sum_{i<j} \phi(\vec{R}_j - \vec{R}_i) \quad . \qquad (5.66)$$

Then

$$\vec{\nabla}_i \vec{\nabla}_j \Phi = \vec{\nabla}_i \cdot \vec{\nabla}_j \phi_{ij} = -\vec{\nabla}^2_{ij} \phi_{ij} \quad \text{for } i \neq j \qquad (5.67)$$

and

$$\vec{\nabla}^2_i \Phi = \sum_{k \neq i} \vec{\nabla}^2_i \phi_{ki} = \sum_{k \neq i} \vec{\nabla}^2_{ki} \phi_{ki} \quad .$$

Thus

$$\Gamma_N = \sum_{i \neq j} \sum \frac{\sin^2 \vec{k} \cdot \vec{R}_i}{k^2} \nabla^2 \phi_{ij} - \sum_{i \neq j} \sum \frac{\sin k \cdot R_i \sin k \cdot R_j}{k^2} \nabla^2 \phi_{ij} \ ,$$

or

$$\Gamma_N = \frac{1}{2} \sum_{i<j} \sum \nabla^2 \phi_{ij} (\sin \vec{k} \cdot \vec{R}_i - \sin \vec{k} \cdot \vec{R}_j)^2 / k^2 \quad . \qquad (5.68)$$

And since

$$\sin \alpha - \sin \beta = 2 \sin \frac{\alpha - \beta}{2} \cos \frac{\alpha + \beta}{2} \quad ,$$

we find

$$\sin \vec{k} \cdot \vec{R}_i - \sin \vec{k} \cdot \vec{R}_j = 2 \sin [\frac{\vec{k} \cdot (\vec{R}_i - \vec{R}_j)}{2}] \cos [\frac{\vec{k} \cdot (\vec{R}_i + \vec{R}_j)}{2}] .$$

And letting

$$\Gamma_N = \frac{1}{2} \sum_{i \neq j} \gamma_{ij} \ .$$

Now

$$|\gamma_{ij}| \leq \frac{2}{k^2} |\nabla^2 \phi_{ij}| \sin^2 \frac{\vec{k}}{2} . \vec{R}_{ij}$$

$$\leq \frac{2}{k^2} |\nabla^2 \phi_{ij}| (\frac{k}{2} . R_{ij})^2 \leq \frac{1}{2} |\nabla^2 \phi_{ij}| \frac{k^2}{k^2} R^2_{ij}$$

or

$$|\gamma_{ij}| \leq \frac{1}{2} |\nabla^2 \phi_{ij}| R^2_{ij} \quad . \qquad (5.69)$$

Knowing the behavior of ϕ_{ij} as $R_{ij}\to\infty$ will give us the behavior of γ_{ij}. Also note that Γ_N as a function of $\vec{k}$ is bounded.

Assume

$$\phi(R) \sim \frac{\lambda}{R^n} \quad , \tag{5.70}$$

Then

$$\vec{\nabla}\phi = \frac{d\phi}{dR}\frac{\vec{R}}{R}$$

$$\vec{\nabla}\cdot\vec{\nabla}\phi = \frac{d^2\phi}{dR^2}\frac{\vec{R}}{R}\cdot\frac{\vec{R}}{R} + \frac{d\phi}{dR}\frac{3}{R} - \frac{d\phi}{dR}\frac{R}{R^2}\cdot\frac{\vec{R}}{R}$$

$$= \frac{d^2\phi}{dR^2} + \frac{d\phi}{dR}\{\frac{2}{R}\} \quad . \tag{5.71}$$

Then

$$\nabla^2\phi \sim \frac{\lambda}{R^{n+2}} \; (= \frac{n(n-1)\lambda}{R^{n+2}} - \frac{2n\lambda}{R^{n+2}}) \; , \tag{5.72}$$

and

$$R^2\nabla^2\phi \sim \frac{\lambda}{R^n} \quad .$$

Thus γ_{ij} has the same asymptotic behavior at long distances as ϕ_{ij}.

(v) <u>Summation of $\vec{k}$</u>

We now return to Eq. (5.61) and using (5.63) write

$$S_N(\vec{k}+\vec{K}) \geq \frac{(\vec{k}+\vec{K})^2}{k^2}\,\frac{(\rho_{\vec{K}}-\rho_{2\vec{k}+\vec{K}})^2}{\rho+\gamma_n(\vec{k})} \geq 0 \quad . \tag{5.73}$$

We now sum this inequality over all values of $\vec{k}\neq 0$ after multiplying by a gaussian distribution:

$$G(\vec{q}) = (\frac{\alpha}{\pi})^{3/2}\, e^{-\alpha\vec{q}\cdot\vec{q}} \tag{5.74}$$

$$C\sum_{\vec{q}\neq\vec{K}} S_N(\vec{q})e^{-\alpha q^2} \geq C\sum_{\vec{k}\neq 0}\frac{(\vec{k}+\vec{K})^2}{k^2}\,\frac{(\rho_{\vec{K}}-\rho_{2\vec{k}+\vec{K}})}{\rho+\gamma_N(\vec{k})} \quad . \tag{5.75}$$

If we assume a long-range order in the crystal appropriate to the given space lattice then

$$\rho_{2\vec{k}+\vec{K}} = 0 \quad \text{for } k < \frac{1}{2} K_o \quad , \tag{5.76}$$

where K_0 is the length of the shortest reciprocal lattice wave-vector. Since each term in (5.74) is positive we can write

$$C \sum_{\vec{q}\neq\vec{K}} S_N(\vec{q}) e^{-\alpha q^2} \geq C \sum_{\substack{\vec{k}\neq 0 \\ k<1/2\, K_o}} \frac{(\vec{k}+\vec{K})^2}{k^2} - \frac{\rho_{\vec{K}}^2}{\rho+\gamma_N(k)} \quad . \tag{5.77}$$

By multiplying by $\frac{1}{V}$ and taking the thermodynamic limit we find

$$C \int d\vec{q}\, S(\vec{q}) e^{-\alpha q^2} \geq C \rho_{\vec{K}}^2 \int_{k<\frac{K_o}{2}} d\vec{k} \frac{(\vec{k}+\vec{K})^2}{k^2 [\rho+\gamma(k)]} \quad , \tag{5.78}$$

where

$$\gamma(k) = \lim_{N,V\to\infty} \gamma_N(\vec{k}) \quad , \tag{5.79}$$

But $\int dq\, e^{-\alpha q^2} S(\vec{q}) < \infty$. (5.80)

[See eq. (5.53) and interchange summation and limiting processes.] We also see that in 1 and 2 dimensions

$$\int \frac{d\vec{k}}{k^2} \frac{(k+K)}{\rho+\gamma(\vec{k})} = \infty \text{ if } \vec{K}\neq 0$$

Therefore we conclude that for $\vec{K}\neq 0$, $\rho_{\vec{K}}^2$, must vanish in 1 and 2 dimensions.

G. Final Comments on the Fluid-Solid Transition

Although there is no exact theory supporting a phase transition in a system of hard spheres, the "experimental" evidence is very persuasive and can be outlined as follows:

(1) Bridgman's high-pressure measurements of the melting line convinced him that at sufficiently high pressure molecules of any "shape" would crystalize [35].

(2) O. K. Rice discovered that glass spheres randomly thrown together form a stable random packing of lower density than a close packed solid. This indicates a transition.

(3) Machine experiments indicate a solid-fluid transition and so far provide the only quantitative information for hard spheres in the transition region [19].

(4) Machine experiments repeated for hard disks show the same type of transition as in the rigid sphere system [19].

(5) Also similar results for a Lennard-Jones potential have been obtained in two-dimensions by de Wette, Allen and Hughes [36] and in three dimensions by Hansen and Verlet [37, see also 38].

For more precise details of the techniques used the reader is referred to [19] and [G.10]. The evidence is strong that there exists a first order phase transition from a solid to a fluid like structure in two dimensions.

Concerning the existence or not of a critical point between the liquid and solid phases the reader will find useful a recent paper by Emch, Knops and Verboven [39] where a rigorous basis is provided for Landau's earlier argument for the non-existence of such a critical point [1]. From the experimental point of view such a critical point seems not to be attainable.

References of General Interest (G): The list of general references and references is not meant to be exhaustive, but takes into account those references related to the lecture notes. See also the references listed in Professors Résibois' and Lebowitz' lecture notes.

G.1. "Critical Phenomena": Proceedings of a Conference, Washington (1965) edited by M. S. Green and J. V. Sengers, N.B.S. pub. 273, U.S. Department of Commerce, Washington (1966).

G.2. M. E. FISHER: "The Theory of Equilibrium Critical Phenomena", Rep. Progr. Phys. 30, II, pp. 615-730 (1967).

G.3. P. HELLER: "Experimental Investigations of Critical Phenomena" Rep. Progr. Phys. 30, II, pp. 731-826 (1967).

G.4. P. A. EGELSTAFF AND J. W. RING: "Experimental Data in the Critical Region", in Physics of Simple Liquids, pp. 253-297, edited by Temperley, Rowlinson and Rushbrooke, North Holland, Amsterdam (1968). On general grounds this book is quite interesting.

G.5. L. P. KADANOFF et. al.: Rev. Mod. Phys. 79, 395 (1967).

G.6. D. SETTE: "Research on Critical Phenomena", The Growth Points of Physics, E.P.S. Inaugural Conference, Florence 1968, Il Nuovo Cimento, Numero speciale, vol. I, pp. 403-441 (1969)

G.7. B. L. SMITH: "Critical Point Phenomena", Contemporary Physics 10, pp. 305-329 (1969).

G.8. "Critical Phenomena": Rendiconti S.I.F. "Enrico Fermi" Corso LI, Varenna Sul Lago di Como, edited by M. S. Green, to appear 1970.

G.9. For often referred original literature the reader may be interested in consulting: The Equilibrium Theory of Classical Fluids (a Lecture note and reprint volume) compiled by Frisch and Lebowitz, W. A. Benjamin Inc., N. Y. (1964).

G.10. Useful material and further references can be found in the Proceedings of the I.U.P.A.P. Statistical Mechanics Meeting held at Kyoto (Japan) September 1968, J. Phys. Soc. (Japan) 26 supp. 1969.

References:

[1] L. D. LANDAU AND E. M. LIFSHITZ: Statistical Physics (2nd edition), Addison-Wesley Pub. Co., Reading (1969).

[2] P. EHRENFEST: Commun. Kamerlingh Omnes Lab. Univ. Leiden Suppl. 75b (1933).

[3] J. GRINDLAY: Can. J. Phys. 46, 2253 (1968).

[4] J. L. BIRMAN: "Symmetry Changes, Phase Transitions and Ferroelectricity" in Ferroelectricity (Proceedings of a Symposium-1966) pp. 20-61, edited by Weller, Elsevier Pub. Co., Inc., N. Y. (1967). In this paper the reader will find a detailed analysis of Landau's theory of second order phase transitions.

[5] L. TISZA: "On the General Theory of Phase Transitions" in Phase Transformations in Solids, pp. 1-37, edited by Smoluchowski, Mayer and Weyl, J. Wiley, New York (1951) See also V. Dvorak and V. Janovec, Bull. Acad. Sci. USSR Phys. Ser. 33, 165 (1969).

[6] L. ONSAGER: Phys. Rev. 65, 117 (1966).

[7] J. C. WEELER AND R. B. GRIFFITHS: Phys. Rev. 170, 249(1968)

[8] R. BROUT: Phase Transitions, W. A. Benjamin, New York(1965)

[9] see e.g. T. L. HILL: Statistical Mechanics, McGraw-Hill New York (1956) or 15 below.

[10] see in Fundamental Problems in Statistical Mechanics Pt. 2 compiled by E. G. D. Cohen, North Holland, Amsterdam 1968: N. M. HUGENHOLTZ: "Quantum Mechanics of Infinitely Large Systems", pp. 197-227; and

D. RUELLE: "On the Gibbs Phase Rule", pp. 113-39.

[11] D. RUELLE: Statistical Mechanics, W. A. Benjamin, New York (1969).

[12] see e.g. P. A. EGELSTAFF: An Introduction to the Liquid State, Academic Press, London and New York (1967).

[13] P. W. KASTELEYN: J. Math. Phys. 4, 287 (1963).

[14] C. N. YANG AND T. D. LEE: Phys. Rev. 87, 404 (1952).
T. D. LEE AND C. N. YANG: Phys. Rev. 87, 410 (1952).

[15] A. MUNSTER: Statistical Thermodynamics, vol. I., Springer-Verlag, New York (1968).

[16] M. SUZUKI: J. Math. Phys. 9, 2064 (1968). In this paper the theorem of Yang and Lee has been extended to the ferromagnetic Ising model with arbitrarily mixed spin values of $S_j = \frac{1}{2}$, 1 and $\frac{3}{2}$ including the case of equal spin values as a speical one.

[17] T. ASANO: Phys. Rev. Letters 24, 1409 (1970). In this paper Yang and Lee's results are extended to the anisotropic Heisenberg ferromagnet. See for a lattice gas O. J. HEILMAN: J. Math. Phys. 11, 2701 (1970). For the dilute Ising model, the anisotropic planar model, the anisotropic classical Heisenberg model and the monomer-dimer model, see H. KUNZ 32A, 311 (1970)-Phys. Letters.

[18] P. C. HEMMER, E. H. HAUGUE AND J. O. AASEN: J. Math. Phys. 7, 35 (1966).

[19] B. J. ALDER AND W. G. HOOVER: "Numerical Statistical Mechanics", in ref. G-4 above, pp. 81-113.

[20] W. W. WOOD: "Monte Carlo Studies of Simple Liquid Models" in ref. G-4 above, pp. 116-230.

[21] N. OGITA et. al.: J. Phys. Soc. (Japan) 26, suppl., 145 (1969). The authors have studied 2-D Ising model type systems on the computer and obtained good agreement with

the exact results of Onsager.

[22] M. KAC: "Toward a Unified View on Mathematical Theories of Phase Transitions" in reference 10 above, pp. 71-105.

[23] F. H. REE AND D. A. CHESNUT: J. Chem. Phys. 45, 3983 (1966)

[24] A. BELLEMANS AND R. K. NIGAM: Phys. Rev. Letters 16, 1038 (1966).

[25] M. KAC, G. E. UHLENBECK AND P. C. HEMMER: J. Math. Phys. 4, 216 (1963). See also ref. G.4 above.

[26] J. LEBOWITZ AND O. PENROSE: J. Math. Phys. 7, 98 (1966).

[27] P. W. KASTELEYN: "Phase Transitions", in reference 10 above pp. 30-70.

[28] E. LIEB: "The Solution of the Rys F Model", Phys. Rev. Letters 18, 1046 (1967) and "The Solution of the KDP Model", Phys. Rev. Letters 79, 108 (1967); see also Phys. Rev. 162, 162 (1967).

[29] R. ZWANZIG AND J. I. LAURITZEN, JR.: J. Chem. Phys. 48, 3351 (1968).

[30] See the review article by J. L. LEBOWITZ: Ann. Rev. Phys. Chem. 19, 389 (1968).

[31] N. D. MERMIN: J. Math. Phys. 8, 1061 (1967). For the impossibility of crystal ordering in one and two-dimensional systems see B. I. SADOVNIKOV AND E. M. SOROKINA: Sov. Phys. Dokl. 14, 968 (1970); Indian J. Pure Appl. Phys. 8, 61 (1970); E. M. SOROKINA: Ibid. 8, 64 (1970), Sov. Phys. Dokl. 15, 23 (1970).

[32] H. FALK: Physica 29, 1114 (1963).

[33] A. ISIHARA: J. Phys. A. (Proc. Phys. Soc.) 1, 539 (1968).

[34] R. E. PEIERLS: Helv. Phys. Acta 7, suppl. 2, 81 (1934). Ann. Inst. H. Poincaré 5, 177 (1935).

[35] P. W. BRIDGMAN: See e.g. Rev. Mod. Phys. 18, 1 (1948), where account of earlier work is given.

[36] F. W. DE WETTE, R. E. ALLEN AND D. S. HUGHES: Phys. Lett. 29a, 548 (1969).

[37] J. P. HANSEN AND L. VERLET: Phys. Rev. 184, 151 (1969).

[38] W. G. HOOVER AND F. H. REE: J. Chem. Phys. 47 (1967), J. Chem. Phys. 49 (1968).

[39] G. EMCH, H. J. KNOPS AND E. J. VERBOVEN: J. Math. Phys. 11 1655 (1970).

[40] H. J. MIKESKA AND H. SCHMIDT: J. Low. Temp. Phys. 2, 371 (1970)

DYNAMICAL EFFECTS AT THE CRITICAL POINT IN FLUIDS AND MAGNETS

Pierre Résibois
Universitê Libre de Bruxelles
Belgique

FOREWORD

The study of dynamical phenomena at the critical point is a rapidly growing field, both from the theoretical and the experimental point of view. For this reason, any general report on the subject suffers the risk of being partly out-of-date at the time of publication.

On the other hand, the original literature is often very specialized and hard to follow for the newcomer; there is without doubt the need for a unified presentation, stressing basic ideas rather than technical aspects.

The hope of the author is that the present report, based on a series of lectures given at the School on Statistical Mechanics in Austin, April 1969, will help in partly meeting this need, and will remain sufficiently general so that most of the results presented will not too rapidly become obsolete. In this spirit, the reader should not expect to find here any detailed treatment but rather to obtain some hints which will make the original papers more easily accessible to him.

CHAPTER I. EQUILIBRIUM PROPERTIES AND INTRODUCTION TO DYNAMICAL PHENOMENA

A. Phase Transition

Equilibrium phase transition is a well-known physical phenomenon in which, generally, the two phases are quite distinct and the transition is abrupt. For instance, the vaporization of water, the crystallization of a super-saturated solute are familiar to everyone. From a mathematical point of view, we have in these cases a first order phase transition in which the thermodynamic potentials have discontinuities in their first derivatives; for instance, in a fluid, the volume V is given by:

$$V = \left(\frac{\partial G(p,T)}{\partial p}\right)_T \quad , \tag{1.1}$$

where G(p,T) is the Gibbs free energy and, in the two phase region, one has:

$$V_L \neq V_G \quad , \tag{1.2}$$

where the subscripts L and G respectively denote the liquid and the gas phase.

However, by changing the thermodynamic parameters, the two phases can be made more and more similar until the critical point is reached, where they become identical. Mathematically, the first derivatives of the thermodynamic potentials then remain continuous but the higher order derivatives show divergences.

For example, the termination point of the liquid-gas coexistence curve determines a critical point (p_c,T_c)(Fig. 1a), which also appears in a well-known manner in the p-V graph of the fluid isotherms(Fig.1b).

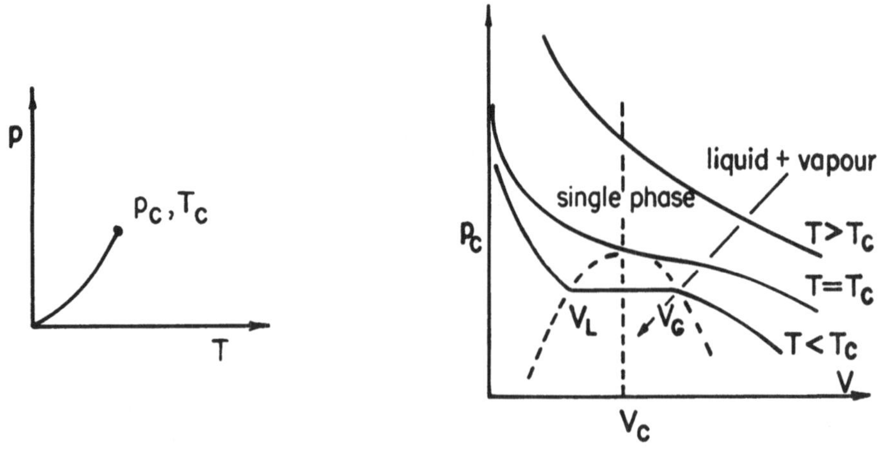

(a) Liquid-gas coexistence curve

(b) p-V isotherms

FIGURE 1. Critical point in a fluid

Another case of interest to us is the spontaneous magnetization M of a ferromagnet (or spontaneous staggered magnetization in an anti-ferromagnet) which occurs in zero applied field H below a critical temperature T_c. The coexistence curve and the M-H isotherms are represented in Fig. 2.

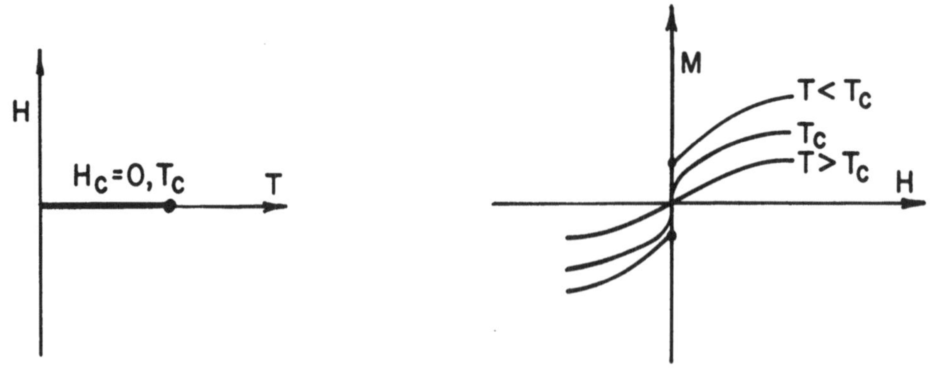

(a) Coexistence curve

(b) M-H isotherms

FIGURE 2. Critical point in a ferromagnet

Other phase transitions are known to occur in binary mixtures, binary alloys, superconductors and superfluid He^4. They will, however, be of no concern to us in these lectures.

By definition, analysis of dynamical effects at (or close to) the critical point is concerned with the time dependent behavior of macroscopic systems which have been very slightly perturbed from their equilibrium state, taken at (or close to) the critical point. For this reason, it is impossible to start a discussion of these effects without a preliminary understanding of the equilibrium properties of macroscopic systems at (or close to) their critical point. This, by itself, is a vast and extremely difficult field which we have no time to discuss in detail. Fortunately, very complete reviews have been written recently on the subject [1-2-3]; this will allow us to be very brief, and to limit ourselves to a sketch of the fundamental ideas needed in the next chapters.

B. Classical Theory and Critical Indices

The basic problem in equilibrium statistical mechanics of critical phenomena is the determination of the asymptotic laws governing the divergences of thermodynamic properties.

The simplest approach (but unfortunately incorrect for realistic systems) to this problem is given by the so-called classical theory (or Van der Waals', or Landau's or Weiss'). Let us consider for definiteness the case of a fluid; let us assume (improperly) that the free energy F(V,T) can be expanded in a Taylor series around V_c and T_c.

$$F(V,T) = F(T,V_c) + \left(\frac{\partial F}{\partial V}\right)_{T,V_c} (V - V_c)$$

$$+ \frac{1}{2}\left(\frac{\partial^2 F}{\partial V^2}\right)_{T,V_c} (V - V_c)^2 + \dots\dots\dots \quad , \tag{1.3}$$

$$\text{with } F(T,V_c) = F(T_c,V_c) + \left(\frac{\partial F}{\partial T}\right)_{T_c,V_c} (T - T_c + \dots) \quad . \tag{1.4}$$

$$\left(\frac{\partial F}{\partial V}\right)_{T,V_c} = \left(\frac{\partial F}{\partial V}\right)_{T_c,V_c} + \left(\frac{\partial}{\partial T}\left(\frac{\partial F}{\partial V}\right)\right)_{T_c,V_c} (T - T_c) + \ldots.. \tag{1.5}$$

etc.

Using the conditions determining the critical point:

$$\left(\frac{\partial p}{\partial T}\right)_{T_c,V_c} = 0; \left(\frac{\partial^2 p}{\partial T^2}\right)_{T_c,V_c} = 0; \left(\frac{\partial^3 p}{\partial T^3}\right)_{T_c,V_c} < 0 \text{ (stability)} \quad , \tag{1.6}$$

and the Maxwell construction for eliminating the metastable and unstable states below T_c, it is a matter of some straightforward thermodynamic calculations to derive the following results:

$T > T_c,\ n = n_c$:	$C_V = \left(\frac{\partial E}{\partial T}\right)_{T,V_c} \propto \varepsilon^{-\alpha}$	$\alpha=0$	(1.7)
	$K_T = -\frac{1}{V_c}\left(\frac{\partial V}{\partial p}\right)_{T,V_c} \propto \varepsilon^{-\gamma}$	$\gamma=1$	(1.8)
$T = T_c,\ n \neq n_c$:	$\lvert p - p_c \rvert \propto (n - n_c)^{\delta}$	$\delta=3$	(1.9)
$T < T_c,\ n = n_c$:	$C_V \propto (-\varepsilon)^{-\alpha'}$	$\alpha'=0$	(1.10)
$n = n_L$ or n_G:	$K_T \propto (-\varepsilon)^{-\gamma'}$	$\gamma'=1$	(1.11)
$n = n_c$:	$(n_L - n_G) \propto (-\varepsilon)^{\beta}$	$\beta=1/2$	(1.12)

Here we have introduced the conventional notation:

$$\varepsilon = (T - T_c)/T_c \quad , \tag{1.13}$$

as well as the so-called critical indices $\alpha, \alpha', \gamma, \gamma', \beta, \delta$ which, crudely speaking, indicate the nature of the divergence of the various thermodynamic properties considered in (1.7,12)(for precise definition, see [1]). Note that, in the present case, $\alpha = \alpha' = 0$ indicate a finite discontinuity in the specific heat at constant volume, C_V. K_T is the isothermal compressibility.

Another quantity of interest, observable by X rays, light, neutrons, is the pair correlation function g(r) defined by:

$$g(r) = \langle \delta n(r) \delta n(0) \rangle / n \quad , \tag{1.14}$$

where <...> denotes an equilibrium average and $\delta n(r)$ expresses the fluctuation of the microscopic density n(r):

$$\delta n(r) = n(r) - \langle n(r) \rangle \quad . \tag{1.15}$$

Using a fairly natural extension of the classical theory [4], one derives the following result. If

$$\tilde{g}(k) = \int dr \exp(ikr) \; g(r), \tag{1.16}$$

one shows that, for $k \to 0$:

$$1 + \tilde{g}(k) = \frac{r_1^{-2}}{K_1^2(T) + k^2} \quad , \tag{1.17}$$

where r_1^{-2} is a smooth function of T around T_c, and where the <u>inverse correlation length</u> $K_1(T)$ has the following behavior:

$$K_1(T) \propto \varepsilon^{\nu,\nu'} \qquad \nu = 1/2 \qquad T > T_c \tag{1.18}$$
$$\nu' = 1/2 \qquad T < T_c \; .$$

Central in the derivation of Eqs. (1.17,18) is the so-called <u>fluctuation theorem</u>:

$$\lim_{k \to 0} (1 + \tilde{g}(k)) = nkT \, K_T \quad . \tag{1.19}$$

This is a classical result of fluctuation theory (see [4]). In the conventional notation for critical indices, we write from (1.17) and (1.18) <u>at $T = T_c$</u>:

$$\lim_{k \to 0} \tilde{g}(k) \propto \frac{1}{k^{2-\eta}} \quad \text{with} \quad \eta = 0 \; . \tag{1.20}$$

This result, which in coordinate space reads:

$$\lim_{r\to\infty} g(r) \propto \frac{1}{r^{1+\eta}} \quad , \tag{1.21}$$

expresses the long range character of the equilibrium correlations at the critical point; it will play a central role in our analysis of dynamical phenomena.

Classical theories can be similarly developed for other critical phenomena. In Table I, we have summarized the critical behavior of the main equilibrium properties of a ferromagnet. In this table, $\tilde{\Gamma}(k)$ is the Fourier transform of the spin correlation function, defined in analogy to (1.17):

$$\Gamma(r) = \langle\delta\vec{s}(r)\,\delta\vec{s}(0)\rangle/s(s+1) \quad . \tag{1.22}$$

In parallel to (1.17), we have now:

$$1 + \tilde{\Gamma}(k) = \frac{r_1^{\,2}}{k^2 + K_1^2(T)} \quad . \tag{1.23}$$

$T > T_c,\ H = 0$:	$C_H = T\left(\frac{\partial S}{\partial T}\right)_{H=0} \propto \varepsilon^{-\alpha}$	$\alpha=0$ (discontinuity)
	$\chi_T = \left(\frac{\partial M}{\partial H}\right)_{H=0,\,T} \propto \varepsilon^{-\gamma}$	$\gamma=1$
	$k_1(T) \propto \varepsilon^{\nu}$	$\nu={}^1/_2$
$T = T_c,\ H \neq 0$:	$\lvert H\rvert \propto \lvert M\rvert^{\delta}$	$\delta=3$
	$\tilde{\Gamma}(k) \propto \frac{1}{k^{2-\eta}}$	$\eta=0$
$T < T_c,\ H = 0$:	$C_H \propto (-\varepsilon)^{-\alpha'}$	$\alpha'=0$ (discontinuity)
	$\chi_T \propto (-\varepsilon)^{-\gamma'}$	$\gamma'=1$
	$K_1(T) \propto (-\varepsilon)^{\nu'}$	$\nu'={}^1/_2$
	$M \propto (-\varepsilon)^{\beta}$	$\beta={}^1/_2$

TABLE I: Critical indices for a ferromagnet and their values as derived from the classical theory.

We shall not dwell further here on these classical theories except to point out that they fail to give a quantitative description of critical phenomena; the incorrect results of these theories derive from the assumed analytic behavior in (1.3). However, it is interesting to note that it has been shown that they are indeed correct in the following (unrealistic) cases:

(a) for an infinite dimensional system, with short range interaction, and

(b) for a 3-dimensional system, with infinite long range forces of the type depicted in Figure 3.

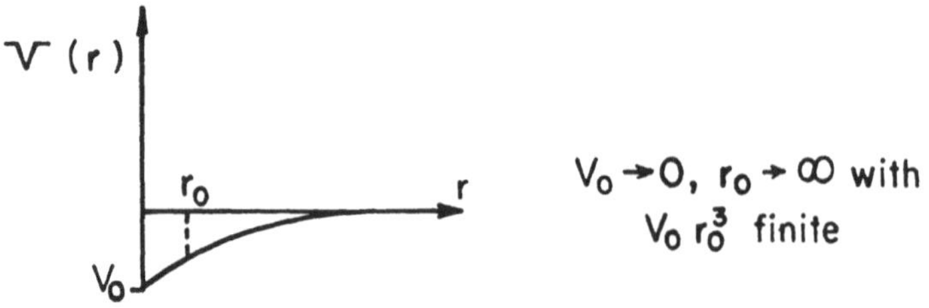

FIGURE 3: Infinite Long Range Potential

In both cases, each molecule interacts with an infinite number of neighbors.

C. Equilibrium Scaling Laws

As we have already mentioned, there is a strong evidence (based on the exact solution of the 2-dimensional Ising model, on computer calculations and on experiments) that although the classical theory is generally wrong, it exhibits most of the qualitative features of real systems.

In order to improve the theoretical situation, a great deal of attention has recently been given to the so-called "scaling laws"; this approach attempts to interrelate the various critical indices.

It is out of place to discuss this theory in detail here. Let us simply mention the central idea of the argument, as it appears in the formulation by Kadanoff [2]: consider a 3-dimensional Ising spin system; because long range effects are dominant in determining critical properties (see (1.21)), it is quite plausible that this Ising system can equally well be described either in terms of individual spins -- with the characteristic parameters ε and H or in terms of interacting macroscopic cells (of length L such that: lattice spacing $\ll L \ll$ correlation length) containing a large number of spins, provided we scale the two parameters ε and H:

$$\varepsilon \to \tilde{\varepsilon}_{cell} = L^y \varepsilon$$
$$H \to \tilde{H}_{cell} = L^x H \qquad . \qquad (1.24)$$

The consequences of this assumption are straightforward and lead to the following equalities (d denotes the number of dimensions):

$$\alpha = \alpha'$$
$$\gamma = \gamma'$$
$$d/y = 2-\alpha = \gamma+2\beta = \beta(\delta+1) \qquad (1.25)$$
$$1/y = \nu = \nu' = 2-\alpha/d$$
$$2x-d = 2-\eta = d\gamma/(2-\alpha) \qquad .$$

These results show that only two critical indices, among nine, are independent. They are exactly satisfied for the 2-dimensional Ising model and are generally well obeyed for all other systems. For instance, the following values are approximately correct for fluids and for the Heisenberg ferromagnet (d = 3)

$$\alpha \simeq \alpha' = 0; \; \beta = 1/3; \; \gamma = \gamma' \simeq 4/3; \; \nu = \nu' \simeq 2/3; \; \eta \simeq 0; \; \delta \simeq 5 \qquad (1.26)$$

and they exactly satisfy Eqs. (1.25).

Nevertheless when data are closely analyzed, small but systematic deviations are found for the scaling laws involving the dimensionality of the system. We refer the reader to [1] and [3] for a detailed analysis of this point.

D. Dynamical Critical Phenomena; Time-Dependent Correlation Functions

Just as equilibrium properties show singularities near T_c, a similar behavior is expected for non-equilibrium parameters. The simplest reason for this is that transport coefficients are closely connected to these equilibrium properties.

To illustrate this point, let us consider the simple case of magnetic diffusion in a ferromagnet : because the total magnetization M is conserved, we have:

$$\partial_t M + \nabla j_M = 0 \quad , \qquad (1.27)$$

where j_M is the magnetization current.

Linear irreversible thermodynamics connects the magnetization current to the magnetic induction B by:

$$j_M = -L \nabla B \quad , \qquad (1.28)$$

where L is the kinetic coefficient of Onsager.

Using the definition of the magnetic susceptibility χ

$$M = \chi B \quad , \qquad (1.29)$$

we obtain from (1.27,28) the diffusion equation:

$$\partial_t M = D \nabla^2 M \quad , \qquad (1.30)$$

where the diffusion coefficient D is

$$D = \chi^{-1} L \quad . \qquad (1.31)$$

The basic idea of the conventional theory is that <u>the Onsager coefficient L is regular through the critical point</u> (because it should depend only on short range effects, insensitive to critical behavior). The singularities of D are then determined by those of χ^{-1}. For instance, because close to T_c,*)

$$\chi_q^{-1} \propto (q^2 + K_1^2) \quad , \qquad (1.32)$$

---

*) See (1.23) and use the equivalent of the fluctuation theorem (1.19) for a ferromagnet, namely: $\lim_{q \to 0} (1 + \tilde{\Gamma}(q)) = \rho K T \chi_q (T)$.

we arrive at:

$$D_q \propto (q^2 + K_1^2) \quad . \tag{1.33}$$

However this result, which is originally due to Van Hove [5], is not verified experimentally; for instance, at T_c, recent neutron scattering experiments indicate that

$$D_q (T = T_c) \propto q^{1/2} \quad . \tag{1.34}$$

The failure of the conventional theory indicates that L is, in fact, not regular at T_c and a detailed microscopic theory is thus needed to investigate its behavior.

All the attempts in this direction have been based on the analysis of time-dependent correlation functions which we write generally:

$$C^A(r;t) = \langle \delta A(r;t) \delta A(0;0) \rangle \quad , \tag{1.35}$$

where the bracket <...> again denotes an equilibrium average and where A(r,t) is some microscopic Heisenberg operator whose fluctuation is denoted by:

$$\delta A = A(r;t) - \langle A(r;t) \rangle \quad . \tag{1.36}$$

The importance of (1.35) is based on the fact that all available experimental results are finally connected to a measurement of (1.35). Let us cite two cases:

a) Light or neutron inelastic scattering experiments:

As was shown by Van Hove, the inelastic cross section $\sigma_q(\omega)$ for momentum transfer q and energy transfer ω is such that:

$$\sigma_q(\omega) \propto \begin{matrix} \text{Re } \tilde{G}(q,\omega) & \text{(fluid)} \\ \text{Re } \tilde{\Gamma}(q,\omega) & \text{(magnet)}, \end{matrix} \tag{1.37}$$

where $\tilde{G}(q,\omega)$ and $\tilde{\Gamma}(q,\omega)$ respectively denote the double Fourier transform of the density-density correlation and of the spin-spin correlation:

$$\tilde{G}(q,\omega) = \int_0^\infty e^{-i\omega t} dt \int e^{iqr} dr \langle \delta n(r,t) \delta n(0,0) \rangle \tag{1.38a}$$

$$G(r,t) \equiv C^{(n)}(r,t)$$

$$\tilde{\Gamma}(q,\omega) = \int_0^\infty e^{-i\omega t} dt \sum_R e^{iqR} \langle \delta \vec{s}(R,t) \delta \vec{s}(0,0) \rangle \tag{1.38b}$$

$$\Gamma(R,t) \equiv C^{(s)}(R,t) \quad .$$

In (1.38b), the sum extends over the lattice points R where the spins are located. The reader may consult reference [5] for a proof of these results.

b) Transport coefficients:

Irreversible statistical mechanics leads to the following expression for a transport coefficient O_i (see [7] and references quoted therein)

$$O_i = \frac{1}{\Omega kT} \int_0^\infty dt \langle \delta J_i(t) \delta J_i(0) \rangle \quad , \tag{1.39}$$

where J_i is the microscopic flux corresponding to the transport coefficient O_i.

For example, the shear viscosity in a simple fluid is given by:

$$\eta = \frac{1}{\Omega kT} \int_0^\infty dt \langle J^{xy}(t) J^{xy}(0) \rangle \quad , \tag{1.40}$$

with the momentum flux:

$$J^{xy} = \sum_i \frac{p_i^x p_i^y}{m} - \frac{1}{2} \sum_{i \neq j} r_{ij}^x \frac{\partial V}{\partial r_{ij}^y} \quad . \tag{1.41}$$

E. Outline of the Course

In the following lectures, we shall review various theories which have been proposed to analyze correlation functions of the type (1.35) close to T_c, both for fluids and for ferromagnets.

In the next chapter, we start with the conventional theory for the density-density fluctuations in a fluid. This theory, although based on considerations analogous to that leading to the incorrect result (1.32), has been more successful for fluids than for magnetic systems. Moreover, it has found interesting - although generally inconsistent - applications in the calculation of transport coefficients themselves. In Chapter III, we review a completely different, but also phenomenological, formulation: the so-called dynamical scaling laws [8], which furnish simple and definite predictions of the critical behavior of dynamical quantities.

Chapter IV is devoted to a more microscopic treatment of critical transport theory in fluids: the theory of Kadanoff and Swift [9] is presently the only attempt to bring non-equilibrium problems to the same level of precision as equilibrium problems; yet many assumptions of these authors should be made explicit before this impressive piece of work can be considered as fully understood.

Finally, in the fifth lecture, we outline a completely microscopic model theory developed by the author and M. De Leener [10] in the case of spin systems; this theory graphically analyzes the Weiss model of a ferromagnet (where the number of neighbors Z tends to infinity). Thus, in view of the remark at the end of paragraph I-B, it is the microscopic dynamical analog of the classical equilibrium theory. Although this theory is very precise (in the sense that it allows a first principle calculation of the spectral function $\tilde{\Gamma}(q,\omega)$), a strong assumption has yet to be done when comparing its conclusions to experimental results.

CHAPTER II. A PHENOMENOLOGICAL APPROACH: HYDRODYNAMICAL DESCRIPTION OF DENSITY-DENSITY FLUCTUATIONS IN A FLUID

A. Linearized Hydrodynamical Description

The conserved macroscopic quantities in a fluid are the particle density $n(r,t)$, the momentum density $g(r,t)$ and the energy density $\bar{\varepsilon}(r,t)$ (involving both the internal energy $\varepsilon(r,t)$ and the kinetic energy $\frac{n}{2}mv^2$). The corresponding conservation equations are (see [7]):*

$$\partial_t n(r,t) + \nabla g(r,t)/m = 0 \qquad \text{(a)}$$

$$\partial_t g(r,t) + \nabla\tau(r,t) = 0 \qquad \text{(b)} \qquad (2.1)$$

$$\partial_t \bar{\varepsilon}(r,t) + \nabla j^{\varepsilon}(r,t) = 0 \qquad \text{(c)} ,$$

where $\tau(r,t)$ is the stress tensor and $j^{\varepsilon}(r,t)$ the energy flow.

In order to obtain from (2.1) a closed system of equations, we use:

1) the phenomenological assumptions for the currents in real fluids (Navier-Stokes equations)
2) the assumption that thermodynamic quantities are,out of equilibrium,interrelated in the same way as at equilibrium.
3) a linearization of the resulting equations around absolute equilibrium.

Using the following definition for the velocity field $v(r,t)$:

$$g(r,t) = n(r,t)mv(r,t) , \qquad (2.2)$$

we obtain from assumptions 1) and 3)

---

* We do not write explicitly the (obvious) tensorial character of the various quantities appearing in this chapter, except when confusion is possible.

$$g(r,t) = nmv(r,t)$$

$$\tau_{ij}(r,t) = p(r,t)\delta^{kr}_{ij} - \eta\left(\frac{\partial v_i(r,t)}{\partial r_j} + \frac{\partial v_j(r,t)}{\partial r_i}\right) - \delta^{kr}_{ij}\left(\zeta - \frac{2}{3}\eta\right)\nabla v(r,t) \qquad (2.3)$$

$$\bar{\varepsilon}(r,t) = \varepsilon(r,t)$$

$$j^{\varepsilon} = hv(r,t) - \lambda\nabla T(r,t) \quad .$$

In these equations, $p(r,t)$ represents the pressure tensor; η, ζ, and λ respectively are the coefficients of shear viscosity, of bulk viscosity and of thermal conductivity; $h = p + \varepsilon$ is the enthalpy density and $T(r,t)$ is the local temperature. We have also used the convention that where the (r,t) dependence of a given variable is not indicated, its absolute equilibrium value should be taken.

We also write:

$$\begin{aligned} n(r,t) &= n + \delta n(r,t) \\ \varepsilon(r,t) &= \varepsilon + \delta\varepsilon(r,t) \; , \end{aligned} \qquad (2.4)$$

and we notice that the second principle of thermodynamics can be written as:

$$\begin{aligned} dS = \frac{dQ}{T} &= \frac{1}{T}(dE + pdV) \\ &= \frac{1}{T}(d(\varepsilon V) + pdV) \\ &= \frac{1}{T}(Vd\varepsilon + (p+\varepsilon)dV) . \end{aligned}$$

We then get for the entropy density fluctuation:

$$ds = \frac{dS}{V} = \frac{1}{T}(d\varepsilon - (p+\varepsilon)dn) \quad , \qquad (2.5)$$

an equation which, according to assumption 2), may be used for the non-equilibrium quantities $\delta s(r,t)$, $\delta\varepsilon(r,t)$ and $\delta n(r,t)$.

Combining Eqs. (2.1,3,4,5), one obtains:

$$\partial_t \delta n(r,t)+n\nabla v(r,t) = 0 \qquad (a)$$

$$\partial_t v(r,t)+\frac{1}{nm}\nabla\delta p(r,t)-\frac{\eta}{nm}\nabla^2 v(r,t)-\frac{(\zeta+{}^{\eta}/_{3})}{nm}\nabla(\nabla v(r,t)) = 0 \qquad (b) \qquad (2.6)$$

$$\partial_t \delta S(r,t)-\frac{\lambda}{T}\nabla^2\delta T(r,t) = 0 \quad , \qquad (c)$$

which will form a closed system of equations provided we use the thermodynamic relations:

$$\delta p(r,t) = \left(\frac{\partial p}{\partial n}\right)_s \delta n(r,t) + \left(\frac{\partial p}{\partial S}\right)_n V\delta s(r,t) \qquad (2.7)$$

$$\delta T(r,t) = \left(\frac{\partial T}{\partial n}\right)_s \delta n(r,t) + \left(\frac{\partial T}{\partial S}\right)_n V\delta s(r,t) . \qquad (2.8)$$

Finally, it is convenient to eliminate the velocity field from Eqs. (2.6). One takes the divergence of Eq. (2.6b):

$$\partial_t(\nabla v(r,t))+\frac{1}{nm}\nabla^2\delta p(r,t)-\frac{\eta}{nm}\nabla^2(\nabla v(r,t))-\frac{(\zeta+{}^{1}/_{3}\eta)}{nm}\nabla^2(\nabla v(r,t)) = 0 \qquad (2.9)$$

and the divergence $(\nabla v(r,t))$ is expressed through the continuity equation (1.6a). We then get:

$$\left(-m\frac{\partial^2}{\partial t^2} + mD_1\frac{\partial}{\partial t}\nabla^2\right)\delta n(r,t)+mc^2\nabla^2\delta n(r,t)+\frac{T}{mnC_v}\left(\frac{\partial p}{\partial T}\right)_v \nabla^2\delta s(r,t) = 0 \qquad (2.10a)$$

$$\frac{\partial}{\partial t}\delta s(r,t)-D_T\frac{C_p}{C_v}\frac{1}{n}\left(\frac{\partial p}{\partial T}\right)_v\nabla^2\delta n(r,t)-\frac{C_p}{C_v}D_T\nabla^2\delta s(r,t) = 0 \qquad (2.10b)$$

In arriving at (2.10),the following definitions and thermodynamic relations have been used:

$$D_1 = ({}^{4}\eta/3 + \zeta)/nm$$

$$D_T = \lambda\frac{V}{T}\left(\frac{\partial T}{\partial S}\right)_p = \frac{\lambda}{mnC_p} \quad \text{(thermal diffusivity)}$$

$$mc^2 = \left(\frac{\partial p}{\partial S}\right)_n = \frac{C_p}{C_v}\left(\frac{\partial p}{\partial n}\right)_T \tag{2.11}$$

$$\lambda\frac{V}{T}\left(\frac{\partial T}{\partial S}\right)_n = D_T\,\frac{C_p}{C_v}$$

$$\frac{V}{T}\left(\frac{\partial p}{\partial S}\right)_n = \frac{1}{mnC_v}\left(\frac{\partial p}{\partial T}\right)_v$$

$$\left(\frac{\partial T}{\partial n}\right)_s = \frac{T}{mn^2C_v}\left(\frac{\partial p}{\partial T}\right)_v \quad .$$

Here C_v and C_p respectively denote the specific heat at constant volume and at constant pressure respectively and c is the velocity of sound.

B. Density-Density Fluctuations: Critical Opalescence

Let us now establish the connection between the results of the preceding section and the density-density time dependent correlation function (1.38a).

We first assume that we want to calculate $G(r,t)$ only for long times and large separations. Under these circumstances, we make the assumption that the microscopic density operator $\delta n(r,t)$ which appears in the definition of $G(r,t)$ can be replaced by the smeared out local macroscopic density fluctuation which obeys Eq. (2.10).

In order to get an equation for $G(r,t)$ we then need to multiply (2.10 a,b) by $\delta n(0,0)$ and to take an equilibrium average. We trivially obtain the same equations as (2.10) with the replacements:

$$\delta n(r,t) \rightarrow G(r,t) \tag{2.12}$$

$$\delta s(r,t) \rightarrow Q(r,t) \equiv \langle \delta s(r,t)\,\delta n(0,0)\rangle \quad . \tag{2.13}$$

Notice that we have a coupling between the density-density

fluctuation and the entropy-density fluctuation; this will turn out to play a very important role close to T_c.

The system of hydrodynamical equations obtained from (2.10) by the substitution (2.12,13) are solved by Fourier-Laplace transform. With the definition used in (1.38a) we arrive at:

$$[i\omega[-i\omega+D_1k^2] - k^2c^2]\bar{G}(k,\omega) - \frac{T}{m^2nC_V}\left(\frac{\partial p}{\partial T}\right)_V k^2\bar{Q}(k,\omega)$$

$$= (-i\omega+D_1k^2)G(k;t=0) \tag{2.14a}$$

$$[-i\omega+k^2D_T\frac{C_p}{TC_V}]\bar{Q}(k,\omega) + k^2D_T\frac{C_p}{TC_V}\frac{1}{n}\left(\frac{\partial p}{\partial T}\right)_V\bar{G}(k,\omega) = Q(k;t=0). \tag{2.14b}$$

This linear inhomogeneous system of equations is of the form:

$$\begin{aligned} \alpha\bar{G}(k,\omega) + \beta\bar{Q}(k,\omega) &= a \\ \gamma\bar{G}(k,\omega) + \delta\bar{Q}(k,\omega) &= b \quad , \end{aligned} \tag{2.15}$$

which leads immediately to:

$$\bar{G}(k,\omega) = \frac{a\delta-b\beta}{\alpha\delta-\beta\gamma} \quad . \tag{2.16}$$

When explicitly written, this result is very awkward unless it is explicitly taken into account that only the long wave length limit $k\to0$ is of interest. This allows us:

1) to neglect terms of order $\left(\frac{D_Tk}{c}\right)^2$, $\left(\frac{D_1k}{c}\right)^2 \ll 1$

2) to replace the initial correlations $G(k,t=0)$ and $Q(k,t=0)$ by their $k=0$ thermodynamic limits. Indeed we have (see (1.19)):

$$\lim_{k\to0} G(k,t=0) = n^2kTK_T \quad , \tag{2.17}$$

and similarly (see [4]):

$$\lim_{k\to0} Q(k,t=0) = kT\left(\frac{\partial n}{\partial T}\right)_p \quad . \tag{2.18}$$

These two assumptions permit to rewrite (2.16) in the simple form:

$$\mathrm{Re}\ \bar{G}(k,\omega) = \frac{n^2 kT K_T}{\pi}\left[\left(1-\frac{C_V}{C_p}\right)\frac{D_T k^2}{\omega^2+(D_T k^2)^2} + \frac{C_V}{C_p}\frac{2\omega^2 k^2 \Gamma}{(\omega^2-C^2k^2)^2+(\omega k^2 2\Gamma)^2}\right] \tag{2.19}$$

with: $\Gamma = {}^1/_2(D_T({}^{C_p}/_{C_V}-1)+D_1)$ <u>sound absorption coefficient</u> (2.20)

Equation (2.19) describes (see Fig. 4)

1) two peaks around $\delta=\pm ck$ corresponding to <u>adiabatic sound wave propagation</u> (<u>Brillouin scattering</u>); their width is of order Γk^2.

2) a central diffusive peak (<u>Rayleigh scattering</u>) related to the non-adiabatic part of the density fluctuation (coupling $(\delta n,\delta s)$)

The respective weights of these peaks is $\frac{C_V}{C_p}$ and $(1-\frac{C_V}{C_p})$, a result due to Landau and Placzek.

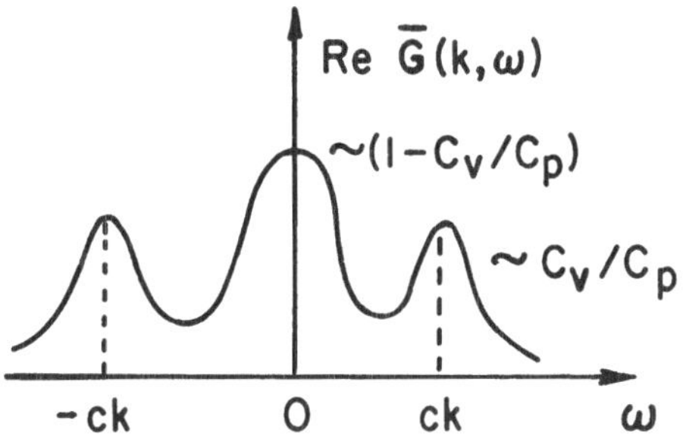

<u>FIGURE 4. Density-density fluctuations in a fluid</u>

The above result is of course valid in the whole fluid region. However, near T_c it gives rise to the phenomenon of <u>critical opalescence</u>.

Indeed, from the thermodynamic relation:

$$C_p = C_V + nmT\left(\frac{\partial p}{\partial T}\right)_V^2 K_T \quad , \tag{2.21}$$

where now (see (1.7,8)

$$C_v \sim \ln\varepsilon = \varepsilon^0$$

$$K_T \sim \varepsilon^{-\gamma} \quad \gamma>0 \tag{2.22}$$

$$\left(\frac{\partial p}{\partial T}\right)_n \sim \text{non singular ,}$$

we see that C_p diverges much more strongly than C_v when approaching the critical point: the diffusive mode then becomes dominant.

Moreover, if we assume that λ remains regular through T_c, we see from (2.11) that $D_T \propto C_p^{-1}$ tends rapidly to zero: we have a critical narrowing, which is the analogy for a fluid of the result (1.33) obtained previously for a ferromagnet.

C. Further Developments: Application To Other Transport Coefficients

In the above calculation, we have assumed that local and linearized hydrodynamics holds, which is not evident when the correlation length $\xi = K_1^{-1}$ becomes larger than, or of the order of, the hydrodynamical length q^{-1}. Moreover, we have assumed that the transport coefficients (η, ζ, λ) are regular through T_c.

These problems have received attention in works by Fixman and coworkers, Kawasaki, Felderhoff, Zwanzig and coworkers...(see [9] for references).

(a) Non-local effects:

In order to discuss this point, it is slightly simpler to work in the variables $\delta n(r,t)$ and $\delta T(r,t)$ instead of $\delta n(r,t)$, $\delta s(r,t)$.

We write now, in place of (2.7):

$$\delta p(r,t) = \left(\frac{\partial p}{\partial n}\right)_T \delta n(r,t) + \left(\frac{\partial p}{\partial T}\right)_n \delta T(r,t). \tag{2.23}$$

Because $\left(\frac{\partial p}{\partial n}\right)_T \propto K_T^{-1} \propto \varepsilon^{\gamma} \to 0$, it is reasonable to take non-local effects into account in the density term , when approaching T_c. We write thus:

$$\delta p(r,t)|_{\delta T=0} = \int \pi(r-r')\delta n(r',t)dr' , \tag{2.24}$$

or in Fourier components,

$$\delta p_k(t) = \pi_k \ \delta n_k(t) \quad . \tag{2.25}$$

The local approximation used earlier is equivalent to (see (1.17,19)):

$$\begin{aligned} \pi_k &\simeq \pi_o \\ &= \left(\frac{\partial p}{\partial n}\right)_T = \frac{1}{nK_T} = kTr_1{}^2K_1{}^2 \underset{T\to T_c}{\to} 0 \quad . \end{aligned} \tag{2.26}$$

From this result it is quite natural to introduce non-local effects by writing instead of (2.26):

$$\begin{aligned} \pi_k &= kTr_1{}^2 \quad (K_1{}^2+k^2 +\ldots) \\ &= \left(\frac{\partial p}{\partial n}\right)_T (1+k^2/K_1^2) \quad . \end{aligned} \tag{2.27}$$

We have thus, in coordinate space:

$$\delta p(r,t)|_{\delta T=0} = \left(\frac{\partial p}{\partial n}\right)_T \left[\delta n - \frac{1}{K_1^2} \nabla^2 \delta n\right] . \tag{2.28}$$

This type of argument can be made more rigorous and can then be incorporated into the scheme developed above. The qualitative conclusions regarding the behavior of $\overline{G}(k,\omega)$ remain unaffected although quantitatively, many weaknesses persist:

1) Is the Taylor series expansion (2.27) more justified than in the classical equilibrium theory?

2) Are the other thermodynamic derivatives more regular than K_T or, at least, do they give smaller corrections than those predicted

on the basis of (2.21)?

3) Are the transport coefficients η, ζ, λ themselves regular functions of q, ω, ε?

These questions will be partly answered in lecture IV, when discussing the theory of Kadanoff and Swift.

(b) Singularities in transport coefficients:

The density-density fluctuation G(r,t), besides its intrinsic interest, also plays an important role in the theory of critical singularities of transport coefficients, as developed by Kawasaki, Zwanzig and Mountain...after the pioneering work of Fixman.

Consider for instance the correlation function formula (1.40) for the shear viscosity η.

If there is a critical behavior of η close to T_c, it is expected to derive from long distance and long time correlations between the fluxes $J^{xy}(t)$ and $J^{xy}(0)$; moreover, it is generally believed that the only independent macroscopic variable which shows important fluctuations is the density fluctuation $\delta n(r,t)$. If these arguments are correct, macroscopic fluctuations in $J^{xy}(t)$ should come from a functional dependence:

$$J^{xy}(t) = J(\delta n(r,t)) \cdot \qquad (2.29)$$

Simple symmetry arguments lead more precisely to

$$J^{xy}(t) = \alpha \frac{\partial}{\partial x} \delta n(r,t) \frac{\partial}{\partial y} \delta n(r,t) + \ldots . \qquad (2.30)$$

Inserting (2.30) into (1.40), one readily obtains the critical behavior of η in terms of G(r,t).

The appealing results deduced from this approach will however not be discussed here anymore because they are presently superseded by the more powerful work of Kadanoff and Swift.

CHAPTER III. ANOTHER PHENOMENOLOGICAL POINT OF VIEW: DYNAMICAL SCALING

A. Formulation

Let us consider the time dependent correlation function $C^A(r,t)$ (see (1.35)) and its Fourier transform $\tilde{C}^A(k,\omega)$ defined by:

$$C^A(r,t) = \int_{-\infty}^{+\infty} \frac{d\omega}{2\pi} \int \frac{d^3k}{8\pi^3} \exp i(kr-\omega t)\tilde{C}^A(k,\omega) . \tag{3.1}$$

For t=0, it reduces to the static function:

$$C^A(r) = \langle \delta A(r,0)\delta A(0,0)\rangle \quad , \tag{3.2}$$

whose Fourier transform is defined by:

$$C^A(r) = \int \frac{d^3k}{8\pi^3} \exp(ikr)\ \tilde{C}^A(k) . \tag{3.3}$$

Comparing (1.32) and (3.1,2,3), we arrive at the sum-rule:

$$\tilde{C}^A(k) = \int_{-\infty}^{+\infty} \frac{d\omega}{2\pi}\ \tilde{C}^A(k,\omega) \quad . \tag{3.4}$$

Let us now briefly summarize the formulation of static scaling when $H_\psi=0$*: the long wave length limit of $\tilde{C}^A(k)$ defines a static susceptibility which, by hypothesis, diverges at T_c**:

$$\lim_{k\to 0} \tilde{C}^A(k) = kT\chi^A = \frac{\partial\langle A\rangle}{\partial(\text{conjugate variable})} . \tag{3.5}$$

For instance, we have for the density-density fluctuation (see also (1.19)):

$$\lim_{k\to 0} C^{(n)}(k) = \int dr\ C^{(n)}(r) = n^2kT K_T \quad . \tag{3.6}$$

---

*In general H_ψ denotes the variable conjugate to the order parameter ψ; thus in a magnet $\psi\equiv$magnetization and H_ψ=magnetic field, while in a fluid $\psi\equiv(\rho_L-\rho_G)$ and $H_\psi\equiv$(chemical potential-critical chemical potential).

**Otherwise, we are not allowed to apply the scaling assumption!

Eq. (3.5) shows that <u>the divergence of χ^A expresses long range correlations in $C^A(k)$</u>. More precisely, it is assumed that a correlation length $\xi(\equiv K_1^{-1})$ exists, which tends to infinity for $T \to T_c$ and one writes, in the asymptotic limit where r and ξ are much larger than all the molecular parameters of the problem:

$$\tilde{C}^A(r) = \frac{1}{r^x} g^+(r/\xi) \qquad T>T_c \tag{3.7a}$$

$$\tilde{C}^A(r) = \frac{1}{r^{x'}} g^-(r/\xi) \qquad T<T_c \quad . \tag{3.7b}$$

As there is no singularity at $T=T_c$ for finite r, we deduce that:

$$x = x' \tag{3.8}$$

$$g^+(0) = g^-(0) \ , \tag{3.9}$$

and in Fourier space, we get thus, for k, ξ^{-1} much smaller than all molecular parameters:

$$\tilde{C}^A(k) = \frac{1}{k^y} \, g^{\pm}(k\xi) \qquad , \tag{3.10}$$

where $y = d-x$ and $g^+(\infty) = g^-(\infty)$. In these formulas, the superscripts + and - respectively correspond to $T>T_c$ and $T<T_c$.

The generalization of these assumptions to dynamical phenomena goes as follows. We first write $\tilde{C}^A(k,\omega)$ in the following way:*

$$\tilde{C}^A_\xi(k,\omega) = 2\pi \frac{1}{\omega^A_\xi(k)} \tilde{C}^A_\xi(k) f^A_{k,\xi}(\omega/\omega^A_\xi(k)) \ , \tag{3.11}$$

where the conditions:

$$\int_{-\infty}^{+\infty} dx f_{k,\xi}(x) = 1 \tag{3.12}$$

$$\int_{-1}^{+1} dx f_{k,\xi}(x) = {}^1/_2 \quad , \tag{3.13}$$

---

*Notice that, in these formulas, an index ξ is used to explicitly denote the temperature dependence of the various quantities.

unambiguously define the characteristic frequency $\omega_\xi^A(k)$ and the "shape" of the spectral function $f_{k,\xi}^A(x)$ for any given $\tilde{C}_\xi^A(k,\omega)$.

To illustrate these formal definitions, let us consider the case where $\tilde{C}_\xi^A(k,\omega)$ is a Lorentzian:

$$\tilde{C}_\xi^A(k,\omega) = A_{k,\xi} \frac{D_{k,\xi}}{\omega^2+(D_{k,\xi})^2}. \tag{3.14}$$

The sum rule (3.4) leads to:

$$\tilde{C}_\xi^A(k) = A_k \int_{-\infty}^{+\infty} \frac{d\omega}{2\pi} \frac{D_{k,\xi}}{\omega^2+(D_{k,\xi})^2} = \frac{A_k}{2} \quad . \tag{3.15}$$

Inserting this result into (3.11), we get:

$$\frac{2\pi}{\omega_\xi^A(k)} f_{k,\xi}^A(\omega/\omega_\xi^A(k)) = \frac{2D_{k,\xi}}{\omega^2+(D_{k,\xi})^2} \quad . \tag{3.16}$$

Condition (3.13) reads thus:

$$\frac{1}{2} = \int_{-1}^{+1} dx f_{k,\xi}^A(x) = \frac{1}{\pi\omega_\xi^A(k)} \int_{-1}^{+1} dx \frac{D_{k,\xi}}{x^2\omega_\xi^A(k)^2+D_{k,\xi}^2}$$

$$= \frac{1}{\pi} \int_{-1}^{+1} dx \frac{\alpha}{x^2+\alpha^2}$$

$$= \frac{2}{\pi} \operatorname{tg}^{-1} \frac{1}{\alpha} \quad , \tag{3.17}$$

where $\alpha = D_{k,\xi}/\omega_\xi^A(k)$. The solution of (3.17) is obviously $\alpha=1$, i.e.:

$$\omega_\xi^A(k) = D_{k,\xi} \quad , \tag{3.18}$$

while, from (3.16), we deduce simply:

$$f_{k,\xi}^A(x) = \frac{1}{\pi(1+x^2)} \quad , \tag{3.19}$$

which indeed satisfies (3.12).

With these definitions in hand, the dynamical scaling assumption of Halperin and Hohenberg, is then formulated as homogeneity conditions analogous to (3.10):

$$\omega_{\xi}^{A}(k) = k^{z}\Omega(k\xi) \tag{3.20}$$

$$f_{k,\xi}^{A}(x) = f^{A}(x,k\xi) \quad , \tag{3.21}$$

where z is an unknown exponent, to be determined from experiment; similarly, the functions $\Omega(y)$ and $f^{A}(x,y)$ remain unspecified by the theory.

B. Consequences and Limitations

Before going into a detailed analysis of dynamical scaling for magnets and fluids, let us point out some very general consequences of Eqs.(3.20,21). Let us first remark that, both experimentally and theoretically, there are three regions of particular interest in the (k,ξ^{-1}) plane depicted in Fig. 5

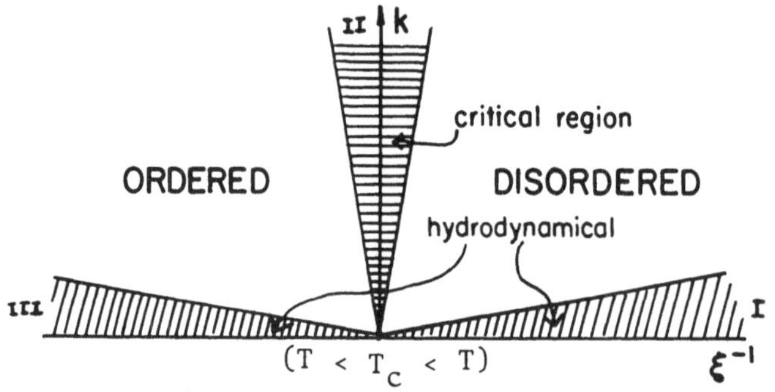

FIGURE 5. The k,ξ^{-1} plane

a) The disordered hydrodynamical region: $k\xi \ll 1$, $T > T_c$ (region I)

b) The non hydrodynamical region: $k\xi \gg 1$, $T \gtrless T_c$ (region II)

c) The ordered hydrodynamical region: $k\xi \ll 1$, $T < T_c$ (region III)

From the homogeneity assumption (3.20,21), supplemented by some very simple physical arguments, it is often possible to extend, by continuity, the results obtained in one of these regions to the two other regions. As it is often much simpler to develop a theoretical analysis in the hydrodynamical regions I or III than in the non hydrodynamical region, Eqs. (3.20,21) then allow one to make predictions about the latter from a knowledge of the former. Moreover, these equations also allow one to compare experimental results taken at different k and ξ and to plot them on universal curves: this introduces a great simplification in the analysis of the data.

We should however keep in mind the limitations which restrict the applicability of Eqs. (3.20,21):

1) As they derive from homogeneity assumptions, which thus far have not been proved, one does not know to which operators A, if any, they do apply*. As we shall see later, this introduces some serious difficulty in the case of fluids.

2) No indication whatsoever is furnished in this theory about the explicit form of the unknown functions $\Omega(x)$ and $f(x,y)$. These are however experimentally available (at least in principle) and should be predicted by a complete theory.

3) As we already mentioned, the most interesting predictions of dynamical scaling require a separate theory for the frequency dependent correlation functions $\tilde{C}^A_\xi(k,\omega)$ in one of the regions depicted in Fig. 5.

C. Application To Fluids and Magnets

1. Fluids: On the basis of very general macroscopic considerations, we have seen in Chapter II that, for $T \to T_c$, the density-density fluctuation satisfies the following equation (see (2.19,21, 22)):

---

* We refer the reader to the original paper of Halperin and Hohenberg [8] for the distinction made by these authors between "extended dynamical scaling" and "restricted dynamical scaling."

$$\mathrm{Re}\ \bar{G}(k,\omega) = \frac{n^2kTK_T}{\pi} \quad \frac{D_Tk^2}{\omega^2+(D_Tk^2)^2} \ . \tag{3.22}$$

In the present notation, this can be written (see (3.6)):

$$\bar{C}^{(n)}(k,\omega) = \bar{C}^{(n)}(k) \ \frac{2D_Tk^2}{\omega^2+(D_Tk^2)^2} \ , \tag{3.23}$$

and, from the argument used after (3.14), we know that:

$$\omega_\xi^{(n)}(k) = D_Tk^2 \tag{3.24}$$

$$f^{(n)}(x) = \frac{1}{\pi} \ \frac{1}{1+x^2} \ . \tag{3.25}$$

Let us suppose that the thermal conductivity λ near T_c diverges as:

$$\begin{aligned} &\lambda \propto |\varepsilon|^{-\ell} \qquad && (T>T_c) \\ &\lambda \propto |\varepsilon|^{-\ell'} \qquad && (T<T_c) \ . \end{aligned} \tag{3.26}$$

From (2.11), (2.21), (1.11) and (1.18) we have then:

$$\omega_\xi^{(n)}(k) \propto \frac{\lambda}{C_p} k^2 \propto k^2 \xi^{(\ell'-\gamma')/\nu'} = k^{2-(\ell'-\gamma')/\nu'} (k\xi)^{(\ell'-\gamma')/\nu'} \quad (T<T_c) \tag{3.27a}$$

and

$$\omega_\xi^{(n)}(k) \propto \frac{\lambda}{C_p} k^2 \propto k^2 \xi^{(\ell-\gamma)/\nu} \propto k^{2-(\ell-\gamma)/\nu} (k\xi)^{(\ell-\gamma)/\nu} \quad (T>T_c) \ . \tag{3.27b}$$

Because the exponent z in (3.20) is the same for $T<T_c$ and $T>T_c$, we get immediately:

$$z = (\ell-\gamma)/\nu = (\ell'-\gamma')/\nu' \ , \tag{3.28}$$

and if the assumptions of static scaling are used ($\nu=\nu'$, $\gamma=\gamma'$), we deduce:

$$\ell = \ell \ , \tag{3.29}$$

a non trivial although somewhat expected result.

Moreover in the <u>non hydrodynamical region</u> ($k\xi>>1$) we have:

$$\omega_{\xi}^{(n)}(k) \propto k^{z}\Omega(k\xi>>1) \propto k^{z} \propto k^{\,2-(\ell-\gamma)/\nu} \quad , \tag{3.30}$$

provided simply that the $\lim_{x\to\infty} \Omega(x)$ exists. This prediction could be checked experimentally by comparing light scattering ($k\xi<<1$) to neutron scattering ($k\xi>>1$), although, as far as we know, this has not been done yet.

It should be pointed out that, from $\bar{G}(k,\omega)$, nothing can be said about the sound velocity c and sound damping Γ because, close to T_c, the Brillouin peaks are completely dominated by the central Rayleigh peak described by (3.22). We may, however, tentatively consider the longitudinal velocity correlation function $C^{(g\ )}(r,t)$:

$$C^{(g\ell)}(r,t) = \langle g^{\ell}(r,t)g^{\ell}(0,0)\rangle \quad , \tag{3.31}$$

where g^{ℓ} is the component of the velocity field parallel to the gradient.

From the equation of continuity (2.1a), it is easily established that:

$$\bar{C}^{(g\ell)}(k,\omega) = \frac{\omega^2}{k^2}\,\bar{C}^{(n)}(k,\omega) \quad . \tag{3.32}$$

Because the heat mode in $\bar{C}^{(n)}(k,\omega)$ behaves as a Dirac distribution $\delta(\omega)$ for small k (see (3.22)), it does not contribute to (3.32), which is thus <u>entirely controlled by the sound modes</u>, although the intensity of the latter becomes very small.

The characteristic frequency of a sound wave is of course:

$$\omega_{\xi}^{(g\ell)}(k) = ck \quad , \tag{3.33}$$

where c is the velocity of sound. The homogeneity assumptions, Eqs. (3.20,21), tell us that the damping of the sound waves, when expressed in reduced units measured by (3.22), is a homogeneous function of $k\xi$; we have thus:

$$\Gamma k^2 = ck(\Gamma k/c) \quad , \tag{3.34}$$

where the bracketed expression has to be a homogeneous function of $(k\xi)$. This implies that:

$$\Gamma \propto c\xi \quad . \tag{3.35}$$

As we shall see in Chapter IV, the more detailed theory of Kadanoff and Swift has not confirmed this result, except at fairly high frequencies. This partial failure of dynamical scaling, however, is not too surprising if we remember that the static correlation:

$$\bar{C}^{(g\ )}(k) = nmkT \quad , \tag{3.36}$$

is trivially non diverging at the critical point. For such a regular operator, the validity of dynamical scaling is very doubtful indeed (see footnote at the end of Section 3B).

2. Ferromagnets: This case is very favorable because, for $T<T_c$, it is possible to develop a complete theory in the hydrodynamical regime (region III in Fig. 5) showing that weakly damped spin waves exist. We shall not explain this theory here and merely quote the results derived by Halperin and Hohenberg ; the spin wave frequency $\omega(k)$ is given by:

$$\omega(k) = \lambda k^2 \quad ; \tag{3.37}$$

with

$$\lambda \propto \xi^{-1}/M \propto \xi^{-1+\beta'/\nu'} \quad , \tag{3.38}$$

while the spin wave damping:

$$\Gamma(k) = Ak^4 \quad , \tag{3.39}$$

where the explicit form of A is not needed for the present purpose. For small values of k, the characteristic frequency ω_ξ^s is obviously determined by the spin wave frequency $\omega(k)$; using the approximate values $\beta\simeq{}^2/_3$, $\nu'\simeq{}^4/_3$, we get:*

$$\omega^s_\xi(k) = \omega(k) \propto \xi^{-1+{}^1/_2}k^2 = k^{5/2}(k\xi)^{-1/2} . \qquad (3.40)$$

The damping constant A can be predicted by repeating the argument given above for sound waves; we have:

$$\Gamma(k) = Ak^4 = (\frac{Ak^2}{\lambda})\lambda k^2 ,$$

where the bracketed expression has to be homogeneous in $(k\xi)$. We get thus:

$$\frac{Ak^2}{\lambda} \propto \xi^2 k^2$$

or

$$A \propto \lambda\xi^2 \propto \xi^{3/2} . \qquad (3.41)$$

From Eqs.(3.20) and (3.40) we have, quite generally:

$$\omega^s_\xi(k) = k^{5/2}\Omega(k\xi) . \qquad (3.42)$$

If $\Omega(\infty)$ exists, we obtain in the non hydrodynamical regime (region II)

$$\omega^s_\xi(k) \underset{\xi k>>1}{\propto} k^{5/2} . \qquad (3.43)$$

Finally, let us consider the hydrodynamical regime above T_c (region I). Eq. (3.42) still holds of course; moreover, we know that in the disordered region no spin wave exists but a diffusive mode is expected (see the thermodynamic argument leading to (1.30)). We write thus, in region III:

$$\omega^s_\xi(k) \underset{k\xi<<1,\ T<T_c}{=} Dk^2 , \qquad (3.44)$$

---

* More precisely, one has $1-{}^\beta/_{\nu'} = (1-\eta)/2$. However η is very small and deviations due to this factor cannot be observed experimentally in inelastic neutron scattering experiments.

which is compatible with (3.42) only if:

$$D \propto \xi^{-1/2} . \tag{3.45}$$

Eqs. (3.43),(3.45) are remarkably well confirmed experimentally, as well as the corresponding equations which have been derived for antiferromagnets. We shall however not dwell on this point here and refer the reader to refs.[8] and [10], where the original experimental papers are quoted. Let us merely point out that Eqs. (3.43)(3.45) have also been obtained by microscopic model calculations, as will be discussed in Chapter V.

CHAPTER IV. A SEMI-MICROSCOPIC APPROACH: THE THEORY OF KADANOFF AND SWIFT

A. Basic Ideas and Formal Preliminaries

The theory of Kadanoff and Swift is based on a fairly unorthodox formulation of transport phenomena. The principle of the method is however quite clear; let us outline it here.

1) We first formally identify the linearized equations of hydrodynamics with the eigenvalue equations of the Liouville operator associated with those eigenvalues which tend to zero when the wave number q, describing spatial inhomogeneity in the system, also tends to zero. The transport coefficients then appear as proportional to the q^2-contribution of these eigenvalues; they can be obtained by a perturbation analysis of the eigenvalue equations around the ideal fluid taken as zeroth approximation.

2) In this perturbation calculus, one considers only the long range contributions in which a given transport mode is "excited" into pairs (or triplets) of similar transport modes. We have thus a mode-mode coupling analogous for instance to phonon-phonon interactions in solids and, in this scheme, transport coefficients are evaluated in a manner very analogous to that used in the computation of the phonon life-time in solid state theory. For example, the process schematically depicted in Fig. 6 will be computed with the help of the (suitably adapted) "golden rule" of quantum mechanics; in particular, a matrix element will be associated with each of the vertices of Fig. 6.

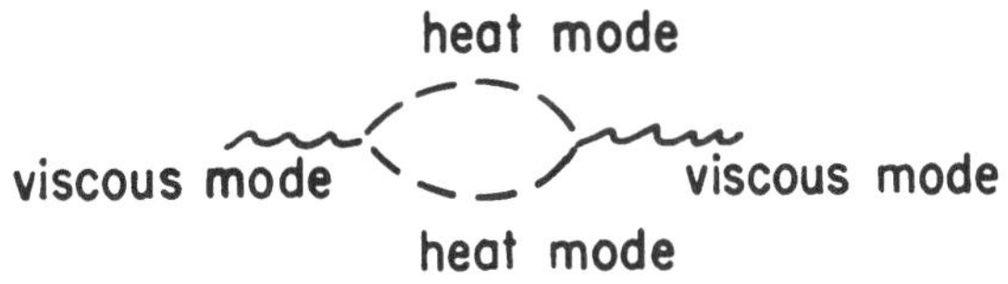

FIGURE 6. Schematic description of the damping of a viscous mode by mode-mode coupling

Let us also stress that the usual (kinetic theory type) contributions to transport coefficients, which come from short range collisions, are neglected in this theory because they should not give any dominant contribution close to T_c.

3) As the matrix elements of the mode-mode coupling processes only involve long wave lengths, they can be determined by thermodynamic arguments, with the help of static scaling laws.

4) As explained in 2), each transport coefficient is determined in terms of other transport coefficients, for instance, in the process depicted in Fig. 6, the shear viscosity is expressed in terms of the thermal conductivity. A systematic analysis of all the possible processes leads to a set of self-consistent equations which are sufficient to determine the critical behavior of these transport coefficients.

Let us now sketch how these ideas can be formulated mathematically We describe the statistical state of the system at time t by a vector $|t\rangle$ in an abstract space. The N-particle distribution function f_N is thus simply:*

$$f_N(r,p,t) = \langle r,p,N|t\rangle \quad . \tag{4.1}$$

In this notation, the Liouville equation reads

$$(\partial_t + L)|t\rangle = 0 \quad , \tag{4.2}$$

where the Liouville operator has the well-known representation in phase-space:

$$\langle r'p'N'|L|r\ p\ N\rangle = \left(\frac{\partial H}{\partial p}\frac{\partial}{\partial r} - \frac{\partial H}{\partial r}\frac{\partial}{\partial p}\right)\langle r'p'N'|r\ p\ N\rangle , \tag{4.3}$$

---

* Except when confusion is possible, we use the notation $r,p \equiv r_1..r_n, p_1...p_n$...to describe the coordinates of the particles in the system.

with the following definition of the scalar product:

$$\langle r'p'N'|r\ p\ N\rangle = \delta^{\kappa r}_{N,N'} \prod_{i=1}^{N} \delta(r_i - r_i') \prod_{i=1}^{N} \delta(p_i - p_i') \ . \qquad (4.4)$$

We denote by $|\rangle$ the grand canonical equilibrium state:

$$\langle r\ p\ N|\ \rangle = \exp -\beta[H- N]/\ \Xi \quad , \qquad (4.5)$$

(Ξ is the grand canonical partition function), and by $\langle\ |$ the "summational state":

$$\langle\ |\ \equiv \sum_N \int dr dp\ \langle r\ p\ N| \quad . \qquad (4.6)$$

While these states are not conjugate they have a "conjugation" property with respect to the Liouville operator:

$$L\ |\ \rangle = 0 \qquad\qquad \langle\ |\ L = 0 \quad . \qquad (4.7)$$

In terms of $|t\rangle$, $|\ \rangle$ and $\langle\ |$, the expectation value of any operator χ_{op} is written:

$$\langle \chi_{op} \rangle_t = \langle\ |\chi_{op}|t\rangle \text{ (non equilibrium average)} \qquad (4.8a)$$

$$\langle \chi_{op} \rangle_{eq} = \langle\ |\chi_{op}|\rangle \text{ (equilibrium average)} \ , \qquad (4.8b)$$

Among the operators χ_{op}, a special role is played by the conserved quantities and their associated currents, namely:

$$\begin{array}{ll} n(r) \text{ (particle density)} & \underline{j}(r) \text{ (particle current)} \\ \underline{g}(r) \text{ (momentum density)} & \underline{\underline{\tau}}(r) \text{ (stress tensor)} \\ \varepsilon(r) \text{ (energy density)} & \underline{j}^{\varepsilon}(r) \text{ (energy current)} \end{array} \qquad (4.9)$$

We shall not need the explicit forms of these operators here; the basic relations between densities and currents are:

$$\begin{aligned} -\ \nabla \underline{j}(r) &= [L,n] \\ -\ \nabla \underline{\underline{\tau}}(r) &= [L,\underline{g}] \\ -\ \nabla \underline{j}^{\varepsilon}(r) &= [L,\varepsilon] \quad . \end{aligned} \qquad (4.10)$$

Moreover, precisely as for the hydrodynamical approach developed in Chapter III, it is convenient to use, instead of the energy density the entropy density operator defined by:

$$S(r) = \frac{1}{T}(\varepsilon(r) - \frac{<\varepsilon+p>}{n} n(r)) \quad , \tag{4.11}$$

with its associated entropy current:

$$j^S(r) = \frac{1}{T}(j^\varepsilon(r) - \frac{<\varepsilon+p>}{n} j(r)) \quad . \tag{4.12}$$

B. Local Equilibrium States and Transport Equations

As in any statistical theory of hydrodynamics, local equilibrium states play an important role in Kadanoff-Swift theory.

In order to construct these states, we first consider the Fourier transform of the five conserved quantities, formally denoted by $a_i'(r)$ ($a_1'(r)=n(r)$; $a_{2,3,4}'(r)=g_{x,y,z}(r)$; $a_5'(r)=\varepsilon(r)$):

$$a_i'(q) = \int d^3r \exp(iqr)\, a_i(r) \quad . \tag{4.13}$$

These operators are not orthonormal in the sense that:

$$<a_i'(q)\, a_j'(-q)>_{eq} \neq \delta_{i,j}\, \delta_{q,q'} \quad . \tag{4.14}$$

It is nevertheless a simple matter to construct orthonormal linear combinations of these $a_j'(q)$; these linear combinations will be denoted by $a_i(q)$ (no superscript) and are explicitly given by:

$$a_1(q) = \frac{S(q)}{(kTmnC_p(q))^{1/2}} \tag{4.15a}$$

$$a_2(q) = (\frac{m}{nkT})^{1/2} c(q) n(q) + \frac{1}{kmn}[\frac{1}{C_v(q)} - \frac{1}{C_p(q)}]^{1/2} S(q) \tag{4.15b}$$

$$a_\gamma(q) = (\frac{1}{mnkT})^{1/2} g_\gamma(q) \quad (\gamma = 3,4,5 \equiv x,y,z) \quad . \tag{4.15c}$$

For finite q, $C_p(q)C_v(-q)$ and c(q) are <u>normalization constants</u> which insure that the orthonormality conditions (we take the volume of the system V=1):

$$\langle a_i(q)a_j(-q)\rangle_{eq} = \delta_{i,j}\delta_{q,q'} \quad , \tag{4.16}$$

are satisfied. Thus for example, we find from (4.15a) and (4.16):

$$C_p(q) \equiv \frac{1}{kTmn} \langle S(q)S(-q)\rangle_{eq} \quad , \tag{4.17}$$

which defines $C_p(q)$.

Yet, the notation is justified because in the long wave length limit, one can show that:

$$\lim_{q\to 0} C_p(q) = C_p, \ \lim_{q\to 0} C_v(q) = C_v, \ \lim_{q\to 0} c(q) = c \quad . \tag{4.18}$$

For example, the first equation of (4.18) is easily deduced from (4.17) by a well-known result of fluctuation theory, analogous to (2.17) and (2.18)(see [4]).

Moreover, as long as we consider values of q such that $q\xi \ll 1$, static scaling tells us that (4.18) can be used in order to estimate the $\varepsilon \equiv \frac{(T-T_c)}{T_c}$ dependence of $C_p(q)$, $C_v(q)$ and c(q).

Another important but subtle consequence of static scaling is that we can also estimate the ε-factors which are introduced by a supplementary s(q) operator in an <u>already fluctuating average value.</u> Consider for example (see(2.18)):

$$\lim \langle S(-q)n(q)\rangle_{eq} = \langle \Delta S \Delta N\rangle = kT\left(\frac{\partial n}{\partial T}\right)_p$$

$$= kT\left(\frac{\partial n}{\partial p}\right)_T \left(\frac{\partial p}{\partial T}\right)_n$$

↖regular.

$$\propto \varepsilon^{-\gamma} \tag{4.19}$$

On the other hand, for $T<T_c$, the liquid or the vapor density obeys (see(1.12)):

$$|\langle n\rangle_{eq} - n_c| \propto (-\varepsilon)^{\beta} \quad . \tag{4.20}$$

Thus adding an operator s(-q) introduces a strongly diverging factor:

$$\underset{q\xi<<1}{s(-q)} \propto \varepsilon^{-\beta-\gamma} \simeq \varepsilon^{-5/3} \quad , \tag{4.21}$$

a result which can be demonstrated quite generally. Similarly, one shows that:

$$a_2(q) \propto \varepsilon^{-1} \quad . \tag{4.22}$$

With the help of these orthonormal operators $a_i(q)$, the local equilibrium states are simply defined by (see(4.5)(4.6)):

$$|i,q\rangle = a_i(q)|\rangle \tag{4.23a}$$

$$\langle i,q| = \langle\ |a_i(-q) . \tag{4.23b}$$

They obviously obey the orthonormality condition:

$$\langle i,q|j,q'\rangle = \delta_{i,j}\delta_{q,q'} \quad . \tag{4.24}$$

Let us now assume that the linear transport modes $|\nu,q\rangle$ of the system are those eigenfunctions of the Liouville operator:*

$$L|\nu,q\rangle = s_{\nu,q}|\nu,q\rangle \quad , \tag{4.25}$$

whose eigenvalues $s_{\nu,q}$ tend to zero for $q\to 0$. We shall not try to justify here this hypothesis; when the consequences of (4.25) will be fully exploited, its validity will appear more clearly.

---

*The translation invariance of L insures that q is a "good" eigennumber of L.

We also extrapolate the ideas of traditional kinetic theory by assuming that these transport modes $|\nu,q\rangle$ are mostly composed of local equilibrium states $|j,q\rangle$ $(j=1,\ldots 5)$.

This latter idea is incorporated into the theory by defining a projection operator P:*

$$P = 1 - \sum_{j=1}^{5} |j,q\rangle\langle j,q| \ , \qquad (4.26)$$

which rejects the local equilibrium components of any state.

We then multiply (4.25) by $\langle j,q|$ and we write:

$$s_{\nu q}\langle j,q|\nu,q\rangle = \langle j,q|L|\nu,q\rangle = \sum_i \langle j,q|L|i,q\rangle\langle iq|\nu,q\rangle + \langle i,q|LP|\nu,q\rangle . \quad (4.27)$$

Similarly, acting with P on (4.25), we get:

$$s_{\nu,q} \ P|\nu,q\rangle = PL|\nu,q\rangle = PLP|\nu,q\rangle + \sum_i PL|j,q\rangle\langle j,q|\nu,q\rangle \ , \qquad (4.28)$$

$$\text{or:} \quad P|\nu,q\rangle = \frac{1}{s_{\nu,q}-PLP} \sum_j PL|j,q\rangle\langle j,q|\nu,q\rangle \ . \qquad (4.29)$$

Inserting this result into (4.27), we obtain a set of equations:

$$\sum_j [s_{\nu,q} \ \delta^{Kr}_{i,j} - L_{ij}(q) - V_{ij}(q,s_\nu)]\langle j,q|\nu,q\rangle = 0 \ , \qquad (4.30)$$

where

$$L_{ij}(q) = \langle i,q|L|j,q\rangle \qquad (4.31)$$

$$V_{ij}(q,s) = \langle i,q|LP \frac{1}{s-PLP} PL|j,q\rangle \ . \qquad (4.32)$$

In principle, these identities do not suffice to determine $s_{q,\nu}$; indeed, the V_{ij} (q,s) involve the complicated operator $(s-PLP)^{-1}$ and this latter is only simple to evaluate if we know the spectrum of L which is of course not the case!

---

*The reader who is not familiar with the use of projection operators in statistical physics is referred, for instance, to Professor Balescu's lectures.

Yet, if one has some approximation to calculate the $V_{ij}(q,s)$, it is tempting to identify the set of equations (4.30) with a statistical description of linear hydrodynamics. Although one cannot prove this result, this identification is made plausible by the fact that the coefficients L_{ij} are identical with the non-dissipative coefficients of linear hydrodynamics while the V_{ij} can be cast into forms which are directly related to the correlation function expressions of transport coefficients (see(1.39)).

(i) Evaluation of L_{ij}: We have:

$$L_{ij} = \langle i,q|L|j,q\rangle = \langle|a_i(-q)L\, a_j(q)|\rangle = \langle|a_i(-q)[L,a_j(q)]|\rangle$$

$$= iq\langle|a_i(-q)j_j(q)|\rangle \quad , \tag{4.33}$$

where (4.7) and (4.10) have been used.

The r.h.s. of eq. (4.33) is a simple equilibrium average, which can be evaluated in the long wave length limit. It can be shown that all L_{ij} vanish except:

$$L_{23} = L_{32} = iq_x\, c(q) \quad , \tag{4.34}$$

if we choose the wave number q parallel to the x-axis.

Thus if we set all $V_{ij}(q,s) = 0$ in Eq. (4.30) we indeed recover the perfect fluid theory (Euler equations) with 3 modes having zero eigenvalues and two modes propagating with sound velocity.

(ii) Evaluation of $V_{ij}(s,q)$:

We have in principle 25 coefficients V_{ij} which, from (4.10), can be rewritten as:

$$V_{ij}(s,q) = -\langle|qj_i(-q)P\frac{1}{s-PLP}Pj_j(q)q|\rangle. \tag{4.35}$$

However, using symmetry properties, it can be shown that only three independent coefficients remain. More precisely, all V_{ij} vanish except:

$$V_{11} = \frac{-q^2\lambda(q,s)}{mnC_p(q)} \qquad V_{22} = \frac{-q^2\lambda(q,s)}{mn}\left[\frac{1}{C_v(q)} - \frac{1}{C_p(q)}\right]$$

$$V_{12} = V_{21} = \frac{-q^2\lambda(q,s)}{mn}\left[\frac{1}{C_p(q)}\left[\frac{1}{C_v(q)} - \frac{1}{C_p(q)}\right]\right]^{1/2} \tag{4.36}$$

$$V_{33} = \frac{-q^2(\zeta + 4\eta/3)}{mn} \qquad V_{44} = V_{55} = \frac{-q^2\eta(q,s)}{mn} ,$$

where

$$q^2\lambda(q,s) = -\frac{1}{k} <| s(-q) LP \frac{1}{s-PLP} PL\, s(q) |> \tag{4.37a}$$

$$q^2(\zeta + 4\eta/3) = -\frac{1}{kT} <| g_x(-q) LP \frac{1}{s-PLP} PL g_x(q) |> \tag{4.37b}$$

$$q^2\eta(q,s) = -\frac{1}{kT} <| g_y(-q) LP \frac{1}{s-PLP} PL g_x(-q) |> \quad . \tag{4.37c}$$

Using again (4.10), we see that the r.h.s. of Eqs. (4.37) involve an autocorrelation between conserved currents; this is precisely the form of the autocorrelation expression for transport coefficients (consider the Laplace transform of (1.39)), and this leads us to identify $\lambda(q,s)$, $\zeta(q,s)$, $\eta(q,s)$ respectively with the thermal conductivity, the bulk viscosity and the shear viscosity. The only difference with the traditional expression for these coefficients is that we have left them in a frequency-and wave-number-dependent form: this is important near the critical point where we expect strong q-and s-dependence.

To summarize this paragraph, we have recovered the structure of the hydrodynamical equations introduced in Chapter II, with, however, a microscopic expression for the q-and s-dependent transport coefficients.

C. Mode-Mode Coupling Approximation and Self-Consistent Determination of Transport Coefficients

In order to evaluate (4.37), we need an explicit form for:

$$\chi = \frac{1}{s-PLP} \quad . \tag{4.38}$$

Formally, we have of course the following representation for the Liouville operator:

$$L = \sum_{\nu} \int \frac{d^3q}{8\pi^3} \, |\nu,q> s_{\nu,q} <\nu,q| \ , \qquad (4.39)$$

where the sum runs over the complete set of eigenstates of L.

However, from (4.26), we know that the projector P discards almost completely the local equilibrium states while it leaves the other states almost untouched. It is thus a good approximation to write, from (4.38) and (4.39), the following formulas:

$$\chi = \int \frac{d^3p}{8\pi^3} \chi_q \qquad (4.40)$$

$$\chi_q \simeq \sum_{\nu'=6}^{\infty} \frac{|\nu',q><\nu',q|}{s-s_{\nu',q}} \ , \qquad (4.41)$$

where, in contrast to (4.39), the sum over ν' is limited to non-hydrodynamical eigenstates.

Let us now take into account that, near T_c, anomalous transport coefficients are expected to come from long time and long wavelength processes. We assume therefore that we may replace the exact states $|\nu',q>$ in (4.41) by the <u>states composed of a product of independent long wavelength transport modes</u>. For instance, we write, for the contribution coming from two independent modes:

$$\Big(\sum_{\nu'} |\nu',q><\nu',q|\Big)_{(2)} \simeq \frac{1}{2} \sum_{\nu_1=1}^{5} \sum_{\nu_2=1}^{5} \int \frac{d^3q'}{8\pi^3} a_{\nu_1}(-q) a_{\nu_2}(-q+q') |><| x$$

$$x a_{\nu_1}(q') a_{\nu_2}(q-q') \ , \qquad (4.42)$$

and we associate the eigenvalue $s_{\nu_1}(q)+s_{\nu_2}(q-q')$ to the component $(\nu_1,q')(\nu_2,q-q')$ of (4.42).

We obtain thus the representation:

$$\chi_q \simeq \frac{1}{2!} \sum_{\nu,\nu'=1}^{5} \int \frac{d^3q'}{8\pi^3} \frac{a_\nu(q')a_{\nu'}(q-q')|><|a_\nu(-q')a_{\nu'}(q-q')}{s-(s_\nu(q'))+s_{\nu'}(q-q'))}$$

$$+ [\quad \text{3-mode contributions}] + \ldots \quad . \tag{4.43}$$

We have obviously not proved Eq. (4.43); the above argument merely indicates its plausibility. In this sense, the present theory remains semi-phenomenological.

Inserting Eqs. (4.38,40,43) into (4.37), we obtain a set of <u>self-consistent equations</u> which allow us to evaluate transport coefficients near T_c*. Indeed, by considering the various possible intermediate states ν,ν' in (4.43), we get equations connecting $\eta(q,s)$, $\lambda(q,s)$, and $\zeta(q,s)$ and involving matrix elements of the type:

$$<|a_\nu(q)a_{\nu'}(q'-q)\ldots.|> \quad . \tag{4.44}$$

These coefficients are equilibrium averages (see(4.36)) and their ε-dependence can be estimated with the help of static scaling by using Eqs. (4.19) and (4.21).

The explicit calculations are however fairly long and will not be reproduced here. We shall merely indicate the type of result which can be obtained.

Considering, for instance, the process depicted in Fig. 6, the following equation is obtained for the coupling viscous flow to heat flow + heat flow:

$$\underset{q\leq\xi^{-1}}{\eta_{TT}(q,s)} \sim \underset{s\leq s_T^*}{kT \frac{nmC_p}{\lambda^*} \xi^{-1}} \quad , \tag{4.45}$$

---

*Clearly, the above reasoning is only valid close to T_c; otherwise, there is no reason to believe that the mode terms (4.43) will give the dominant contributions in the exact expression (4.40).

where $\lambda^* = \lambda(\xi, s_T^*)$ and s_T^* is the solution of the dispersion equation:

$$s_T(\xi) = \lambda(\xi, s_T)\xi^2/mnC_p \quad . \tag{4.46}$$

It should be pointed out that, in (4.45), the ξ-dependence comes in because it is assumed-and verified a posteriori-that the dominant wave-numbers in the integral (4.43) are such that $q \lesssim \xi^{-1}$, while the low-frequency condition $s \lesssim s_T^*$ insures that the denominator in the same formula can become large (when $\nu = \nu' = T$).

Notice that (4.45) immediately leads to an interesting consequence:

$$\eta_{TT}(q,s)\ \lambda^* \propto C_p \xi^{-1} \propto \varepsilon^{-\gamma+\nu} \propto \varepsilon^{-2/3} \ , \tag{4.47}$$

<u>the indication of a necessary divergence in at least one of the transport coefficients</u>.

When the same analysis is applied to the various possible mode-mode couplings, the singular behavior of the various transport coefficients can be determined unambiguously as a function of the frequency. We find essentially four regions with distinct behavior:

$$\begin{array}{ll} \underline{\text{Region I}}: & 0 < s \lesssim s_T^* \propto \varepsilon^2 \\ \underline{\text{Region II}}: & s_T^* \lesssim s \lesssim s_\eta^* \propto \varepsilon^{4/3} \\ \underline{\text{Region III}}: & s_\eta^* \lesssim s \lesssim c\xi^{-1} \propto \varepsilon^{2/3} \\ \underline{\text{Region IV}}: & c\xi^{-1} \lesssim s \end{array} \tag{4.48}$$

The singularities of λ, η and ζ in these various regions, together with the dominant mode-mode couplings, are summarized in Table 2.

As a final remark, let us point out that the sound absorption coefficient Γ (see (4.20)):

$$\Gamma = \frac{1}{2}\frac{\lambda}{mn}\left(\frac{1}{C_p} - \frac{1}{C_V}\right) + \left({}^4\eta/_3 + \zeta\right)/mn \quad , \tag{4.49}$$

can easily be evaluated from this table. We find that $\Gamma \propto c\xi \propto \varepsilon^{-2/3}$

only in Region II and III, while this relation is not verfied in Region I. This is a partial failure of the dynamical scaling prediction (see(4.35)).

MODE	DOMINANT COUPLING	REGION I	REGION II	REGION III	REGION IV
		$s\sim s_T^{*}\sim\varepsilon^2$ (boundary I–II)	$s\sim s_\eta^{*}\sim\varepsilon^{4/3}$ (boundary II–III)	$s\sim c\xi^{-1}\sim\varepsilon^{2/3}$ (boundary III–IV)	
λ	$\lambda + \eta$	$\lambda_1 \propto \varepsilon^{-2/3}$ -----	----→		
	sound sound $+\eta$	$\lambda_2 \propto \varepsilon^{0}$ -----	-----	----→	
η	$\lambda + \lambda$	$\eta_1 \propto \varepsilon^{0}$ ----→			
	sound sound$+\lambda$	$\eta_2 \propto \varepsilon^{0}$ -----	-----	----→	
ζ	$\lambda + \lambda$	$\zeta_1 \propto \varepsilon^{-2}$ ----→			
	sound sound$+\lambda$	-----	$\zeta_2 \propto \varepsilon^{-2/3}$ -----	----→	
λ η ζ	high q-processes (usual kinetic theory type)	-----	regular behavior -----	-----	----→

TABLE 2. Summary of the divergences of transport coefficients close to T_c

CHAPTER V. A MICROSCOPIC MODEL: TIME DEPENDENT FLUCTUATIONS OF THE HEISENBERG SPIN SYSTEM IN THE WEISS LIMIT

A. Formulation of the Problem

As we have already mentioned, neutron scattering furnishes direct measurements of the spin correlation function, which, for an isotropic system above T_c, is simply:

$$\Gamma_{ab}(t) = \langle \bar{s}_a(t)\bar{s}_b(0)\rangle \qquad , \qquad (5.1)$$

because of rotation invariance. We also pointed out in Chapter I that the classical theory (see(1.33)) completely fails to describe the long wavelength and long time behavior of this quantity close to T_c.

Although simple phenomenological arguments like dynamical scaling can be applied to this problem, the simplicity of the Heisenberg Hamiltonian:

$$H = -\sum_{i+j} J_{ij}\ \bar{s}_i\bar{s}_j \qquad , \qquad (5.2)$$

makes it tempting to approach this dynamical problem from a purely microscopic viewpoint. In this lecture, we shall briefly outline an attempt of this sort, developed by M. De Leener and the author.

Let us first briefly introduce the basic formalism and indicate how it is related to the theory of irreversible processes as developed by Prigogine and coworkers.*

Because of isotropy, we may rewrite (5.1) as:

$$\Gamma_{ab}(t) = 3\ \Gamma^{zz}_{ab}(t) = 3\ \mathrm{Tr.}s^z_a\ \exp[-iHt]s^z_b\rho^{eq}\exp[+iHt] \quad , \qquad (5.3)$$

where ρ^{eq} denotes the equilibrium density matrix:

$$\rho^{eq} = [\exp - \beta H]/\mathrm{Tr}[\exp - \beta H] \quad . \qquad (5.4)$$

---

*A presentation of this latter theory can be found in Professor Balescu's lectures.

This can be formally rewritten as:

$$\Gamma_{ab}(t) = 3\mathrm{Tr}\ s_a^z\ \rho^z(t|b) \tag{5.5}$$

where

$$\rho^z(t|b) = \exp[-iHt]\ s_b^z\ \rho^{eq}\exp[iHt]\ . \tag{5.6}$$

This latter quantity is closely related to a density matrix (although $\mathrm{Tr}\ \rho^z(t|b) = 0$); in particular, it obeys the Liouville-Von Neumann equation:

$$i\partial_t\ \rho^z(t|b) = [H,\ \rho^z(t|b)] \equiv \mathcal{H}\rho^z(t|b)\ . \tag{5.7}$$

Let us now construct a representation in which computation with the operator $\mathcal{H}$ is easy to perform. For this aim, we start from the localized spin representation $|\{m_i\}>$:

$$|\{m_i\}> = \prod_{i=1}^{N} |m_i> \tag{5.8}$$

$$s_i^z|m_i> = m_i|m_i>\ , \tag{5.9}$$

and we denote it by the condensed notation $|m>$. We write then for an arbitrary operator A:

$$<m|A|m'> \equiv A_{m-m'}\left(\frac{m+m'}{2}\right) \equiv A_\mu\ (M) \tag{5.10}$$

with

$$\mu = m-m' \qquad M = (m+m')/2\ . \tag{5.11}$$

Here, of course, μ and M denote the entire sets of numbers $\{\mu_i\}$ and $\{M_i\}$. From (5.11), we see that the indices μ_i characterizes the "off-diagonality" of the matrix elements; we shall discuss their physical significance later on.

We can now write the Liouville-Von Neumann equation in this μ-N representation. From (5.7), we get indeed:

$$i\partial_t \langle m|\rho^z(t|b)|m'\rangle = \sum_{m''} (\langle m|H|m''\rangle\langle m''|\rho^z(t|b)|m'\rangle$$

$$- \langle m|\rho^z(t|b)|m''\rangle\langle m''|H|m'\rangle) \quad , \tag{5.12}$$

or, using (5.10):

$$i\partial_t \rho^z_\mu (M;t|b) = \Sigma(H_{\mu-\mu'}(M+\frac{\mu'}{2})\rho^z_{\mu'}(M+\frac{\mu'-\mu}{2};t|b)$$

$$- H_{\mu-\mu'}(M-\frac{\mu'}{2})\rho^z_{\mu'}(M-\frac{\mu'-\mu}{2};t|b)) \quad , \tag{5.13}$$

where, in the first term of the r.h.s. of (5.12), we have set $\mu'=m''-m'$ while, in the second term, we have set $\mu'=m-m''$.

We introduce the displacement operator η such that, acting on an arbitrary function $f(M)$, it leads to:

$$\eta^\mu f(M) = f(M+\mu/2) \quad . \tag{5.14}$$

Eq. (5.13) can then be cast into the following form:

$$i\partial_t \rho^z_\mu (M;t|b) = \sum_{\mu'} \langle\mu|H(M)|\mu'\rangle\rho^z_\mu(M;t|b) \quad , \tag{5.15}$$

where the "Liouville operator" of the problem, $\langle\mu H(M)|\mu'\rangle$, is defined by:

$$\langle\mu|H(M)|\mu'\rangle = \eta^{\mu'} H_{\mu-\mu'}(M)\eta^{-\mu} - \eta^{-\mu'} H_{\mu-\mu'}(M)\eta^{\mu} \quad . \tag{5.16}$$

Eqs. (5.15,16) form the starting point of the analysis of the time dependent spin correlation function $\Gamma_{ab}(t)$. As we shall not perform this analysis in any detail here, we shall not need these explicit equations. There are nevertheless two points worth mentioning:

1) There is a striking similarity between Eq. (5.15) and the Liouville equation for classical gases*. Mutatis mutandis the methods developed in the latter case will be applicable here.

---

*See Professor Balescu's lectures.

2) The Liouville operator $\langle\mu|H(M)|\mu'\rangle$ describes transitions from a "state" $\{\mu'\}$ to a "state" $\{\mu\}$*. This can most conveniently be represented by graphs. However, before sketching this, we should now return briefly to the physical significance of the indices $\{\mu_i\}$.

Let us define the following reduced density matrix:

$$\rho_\mu^z\,(M_a;t|ab) = \mathrm{Tr}'\,\rho^z(t|b) \tag{5.17}$$

where the prime means that the trace is taken over all spins except a.

Using notation (5.10), it is readily seen that:

$$\rho_\mu^z\,(M_a;t|ab) = \sum_{\{M_i\}'} \rho^z_{\mu_a,\{0\}'}\,(\{M\};t|b) \tag{5.18}$$

This one spin density matrix $\rho^z_{\mu_a}$ has only four elements in the case where the spin $|s| = {}^1/_2$:

1) $\rho_o^z\,(M_a,t|ab)$: diagonal elements ($\mu_a=0$ $m_a=m'_a$) with a well defined spin {up: $M_a=+1$; down: $M_a=-1$}

2) $\rho_{-1}^z\,(0,t|ab)$: corresponding to the precession of spin "a" around the z axis in the positive sense.

3) $\rho_{-1}^z\,(0,t|ab)$: corresponding to the precession of spin "a" around the z axis in the negative sense.

Note that in cases 2) and 3), $\mu_a=\pm1$; thus $m_a=m'_a\pm1$ and in both cases, for $|s|={}^1/_2$, we have $M_a=0$.

We see thus that the indices $\{\mu_i\}$ indicate the states of precession of the various spins in the system; for each spin, these states will be represented by graphical symbols, as indicated in Figure 7 (a,b,c).

*It is convenient to read Eq. (5.15) from right to left.

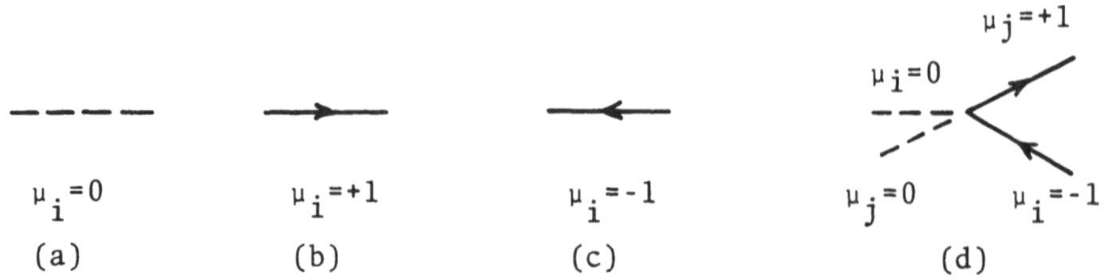

FIGURE 7. Graphical representation of the spin states (a,b,c) and of a typical vertex (d)

If we now come back to the Liouville operator (5.16), we find it very natural to represent the transition $\{\mu'\}\to\{\mu\}$ by a vertex in which the lines representing the interacting spins are modified according to the type of transition; for example, Figure 7d represents the operator:

$$\langle 0_i,\ 0_j | H_{ij} | +1_i, -1_j \rangle \quad . \tag{5.19}$$

We shall not dwell further here on this graph method and we refer the reader to the original papers for details [10].

Let us close this fairly technical section with one more result. From (5.3), (5.10) and (5.17), we can rewrite the spin correlation function in terms of the reduced density matrix $\rho^z_o(M_a,G/ab)$:

$$\Gamma^{zz}_{ab}(t) = \sum_{\{M\}} M_a\, \rho^z_{\{0\}}(M;t|b)$$

$$= \sum_{M_a=\pm 1/2} M_a\, \rho^z_o(M_a,t|ab) \quad . \tag{5.20}$$

Moreover, the conservation of $Tr.\rho^z(t|b)$ immediately implies that ($|s|=1/2$):

$$\rho^z_o(1/2;t|ab) = -\rho^z_o(-1/2,t|ab) \quad .$$

We get finally:

$$\Gamma_{ab}(t) = 3\Gamma^{zz}_{ab}(t) = 3\, \rho^z_0(1/2;t|ab) \quad . \tag{5.21}$$

We see thus that if we can write a kinetic equation for the reduced density matrix $\rho_0^z(^1/_2;t|ab)$, this same equation will immediately apply - apart from trivial constant factors - to the spin correlation function $\Gamma^{ab}(t)$ itself.

B. Kinetic Equations in the Weiss Limit

The analogy of the above formulation with those of particle dynamics makes it obvious that the derivation of kinetic equations given in the latter case can be formally extended with no special difficulty. We shall not reproduce this derivation here and we merely quote the results.

Denoting by $\Gamma_q(t)$ the Fourier transform of $\Gamma_{ab}(t)$(normalized in such a way that $\Gamma_q(0) = 1$), one can show that, indeed, in close resemblance with the particle case, the following equation holds:

$$\partial_t \Gamma_q(t) = -\int_0^t \bar{G}_q(t-t')\Gamma_q(t')dt' \quad , \tag{5.22}$$

where the non markovian kernel $\bar{G}_q(\tau)$ can be defined in graphical terms; we shall not need this definition here. Physically $\bar{G}_q(\tau)d\tau$ gives the probability for a dynamical interaction of the spin density fluctuation $\Gamma_q(t)$ and having a duration τ. Let us also note that, according to (5.21), this kinetic equation directly applies to $\Gamma_q(t)$, which depends on time but not on the spin quantum numbers {M}: this implies that $\bar{G}_q(t)$ is a c-function and not an operator*.

However, when explicit evaluation of $\bar{G}_q(\tau)$ is made, a drastic difference appears with the particle case, which makes Eq. (5.22)

---

*Referring again to Professor Balescu's lectures, let us point out that the kinetic equation (5.22) involves no term describing the "destruction" of the initial correlations: this is due to the particular nature of the initial conditions considered here (see (5.1) for t=0).

meaningless as long as $\bar{G}_q$ is defined by a formal perturbation expansion.

In order to understand this point, let us look naively at $\bar{G}_q$ in the Born approximation in the gas and in the spin cases.

(i) Gas: The second order expression for the transition probability is well known; it has the following form:

$$\bar{G}^{(2)}_{gas}(\tau) \propto \lambda^2 \int d^3q_1 \ldots \exp i(\varepsilon_{k+q_1} + \varepsilon_{k'-q_1} - \varepsilon_k - \varepsilon_{k'})\tau , \qquad (5.23)$$

where q_1 denotes the exchange of momentum between the two incident particles with momenta k and k'. In particular, we have explicitly written in (5.23) the familiar oscillating factor describing the unperturbed motion of the particles between their two interactions taken respectively at times 0 and τ.

When these oscillating factors are integrated over all possible momentum transfers q_1, they lead to a kernel $G^{(2)}_{gas}(\tau)$ which smoothly decays to zero for times larger than than the collision time τ_c:

$$\bar{G}^{(2)}_{gas}(\tau) \underset{\tau \gg \tau_c}{\rightarrow} 0 \quad . \qquad (5.24)$$

This property is depicted in Figure 8a and expresses the fact that, because of their free motion, two particles only interact during a finite time.

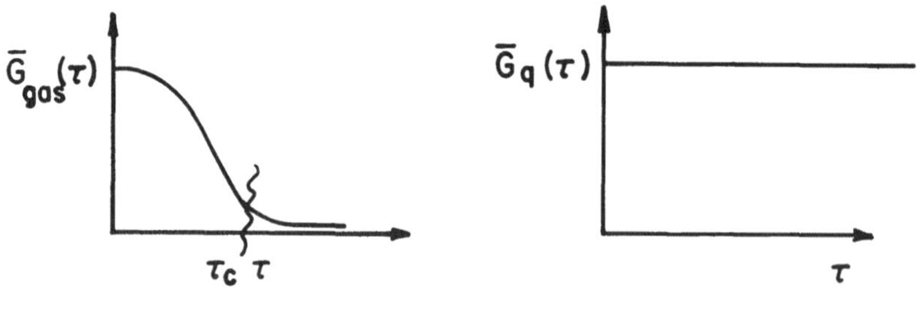

FIGURE 8. Schematic behavior of the kernel in the Born approximation: (a) for a gas (b) for a spin system

(ii) Spins: It is a simple matter to compute $\bar{G}_q^{(2)}(\tau)$ from the Liouville equation (5.15). As we are not interested in the detailed factors, it is easy to understand that, instead of (5.23) we should now have:

$$\bar{G}_q^{(2)}(\tau) \propto \lambda^2 \int d^3q_1 \ldots . 1 \quad . \tag{5.25}$$

Indeed, because we have no unperturbed Hamiltonian in (5.2), the free spins do not propagate between their interactions at times 0 and τ and the oscillating factors appearing in (5.23) are simply replaced by one.

Even after the q_1-integration, we still have:

$$\lim_{\tau\to\infty} G_q^{(2)}(\tau) = C^t \quad . \tag{5.26}$$

This physically meaningless result expresses the fact that, in the Born approximation, the duration of a collision between two neighboring spins is infinite.

However, under these circumstances, we have no right to neglect, even in the roughest approximation, the effect due to the remaining spins. We cannot use the Born approximation to describe the transition probability between two given spins "a" and "b" but we have to take simultaneously into account the effect of the "bath" made of the (N-2) remaining spins. Physically, we expect indeed that the magnetization put initially on "a" and "b" will diffuse over the whole bath; after a certain characteristic diffusion time τ_d, the interaction between "a" and "b", although always present, will become ineffective.

As the function which describes the diffusion of the magnetization is the correlation function $\Gamma_q(t)$ itself, it is reasonable to construct a renormalized kernel $G_q^{(2)}(\tau)$ (no tilde!) in the form:

$$G_q^{(2)}(\tau) \propto \lambda^2 \int d^3q_1 \ldots \Gamma_{q+q_1}(\tau)\, \Gamma_{q_1}(\tau) \quad . \tag{5.27}$$

Note the quadratic dependence on Γ_q, which appears because the magnetization diffuses from each of the spins "a" and "b".

Because $\Gamma_q(\tau)\to 0$ for $\tau \gg \tau_d$, we now have a property analogous to (5.24), namely:

$$G_q^{(2)}(\tau) \underset{\tau \gg \tau_d}{\to} 0 \quad . \tag{5.28}$$

These intuitive arguments can be formalized in theorems which are valid in the <u>Weiss limit</u>, where the number of neighbors Z goes to infinity. In this limit, it has been shown that the whole perturbation series defining $\overline{G}_q(\tau)$ can be rewritten as a functional of $\Gamma_q(\tau)$ itself:

$$\overline{G}_q(\tau) = G_q(\tau|\{\Gamma_{q'}\}) . \tag{5.29}$$

The simplest approximation to (5.29) leads to an equation of the form predicted in (5.27) namely:*

$$G_q^{(2)}(\tau) = \frac{2}{N} \sum_{q_1} J_{q_1} \frac{(J_{q+q_1} - J_{q_1})}{(1-\beta J_{q_1}/2)} \Gamma_{q+q_1}(\tau)\Gamma_{q_1}(\tau)(1-\beta J_q/2) \quad , \tag{5.30}$$

but higher order corrections can also be systematically computed in a kind of cluster expansion:

$$G_q(\tau|\{\Gamma_q\}) = \sum_n G_q^{(2n)}(\tau|\{\Gamma_q\}) \quad , \tag{5.31}$$

which we shall not discuss here.

Thus far, we have said nothing about temperature-effects although these clearly appear through the initial condition $\rho^z(0/b)$ (see (5.6)). They have, however, been explicitly indicated in (5.30) where we find:

---

*This same kernel was obtained by K. Kawasaki, using a completely different method (see [10]).

(a) The fairly trivial factor $(1-\beta J_q/2)$ which comes from our normalization of $\Gamma_q(0)$. Indeed, we have defined $\Gamma_q(0) = 1$ while the Fourier transform of (5.1) at time t=0 is the well known Ornstein-Zernicke correlation function $1/(1-\beta J_q/2)$ [1].

(b) A more subtle dynamical effect, the factor $(1-\beta J_{q_1/2})^{-1}$, which expresses the effect of the initial correlations on the dynamical evolution of the system. Crudely speaking, one can say that the main role of these initial correlations is to introduce temperature dependent interactions:

$$J_q \to J_q^{\text{eff}} = J_q(1-\beta J_q/2)^{-1} \quad , \tag{5.32}$$

The important feature of (5.32) is that it leads to long range effective forces close to T_c. With the notation used in (1.23), we have indeed:

$$\underset{\substack{q\to 0\\ T\to T_c}}{J_q^{\text{eff}}} \propto 1|(q^2 + K_1^2) \underset{T=T_c}{\propto} {}^1/_{q^2} \quad , \tag{5.33}$$

because, in the Weiss limit, the critical temperature is defined by:

$$kT_c = J_{o/2} \quad . \tag{5.34}$$

C. Critical Behavior

In the high temperature limit, Eq. (5.30) becomes:

$$\underset{T>>T_c}{G_q^{(2)}} = \frac{2a^3}{(8\pi^3)} \int d^3q_1\, J_{q_1}(J_{q+q_1} - J_{q_1})\Gamma_{q+q_1}(\tau)\Gamma_{q_1}(\tau) \quad , \tag{5.35}$$

while at T_c, we get:

$$\underset{\substack{T\to T_c\\ q\to 0}}{G_q^{(2)}} = \frac{a^5(kT_c)^2 q^2}{6\pi^3} \int d^3q_1\, \frac{(q-q_1)^2 - q_1^2}{q_1^2}\, \Gamma_{q+q_1}(\tau)\Gamma_{q_1}(\tau) \quad , \tag{5.36}$$

where Eqs. (5.33,34) have been used.

The comparison between (5.35) and (5.36) immediately suggests an anomalous behavior at T_c. Indeed, at high T, we have:

$$G_q^{(2)} \underset{q \to 0}{\propto} q^2 \tag{5.37}$$

indicating a slow decay of $\Gamma_q(t)$ for small q. On the other hand, the dominant q_1-contributions in the kernel come from q_1 inside the Brillouin zone B, for which $\Gamma_{q_1}(t)$ decays much faster. The kernel $G_q^{(2)}(\tau)$ decays thus very rapidly compared to $\Gamma_q(t)$ and we are allowed to replace the non-Markovian equation by a Markovian diffusion equation:

$$\partial_t \Gamma_q(t) \underset{\substack{T \gg T_c \\ q \to 0}}{=} -\int_0^t G_q(\tau|\{\Gamma_{q'}\})\Gamma_q(t-\tau)d\tau$$

$$= -\int_0^\infty G_q(\tau|\{\Gamma_{q'}\})d\tau[\Gamma_q(t)+0(q^2)]$$

$$\simeq D_q^2 \Gamma_q(t) \tag{5.38}$$

where
$$D = -\frac{1}{2}\frac{\partial^2}{\partial q^2}\int_0^\infty G_q(\tau|\{\Gamma_{q'}\})d\tau \tag{5.39}$$

On the contrary, because of the supplementary factor q_1^{-2}, which appears in Eq. (5.36), <u>the important q_1-values in the kernel are themselves of order q at the critical point</u>; there is then no separation between the time-scales of $\Gamma_q(t)$ and of $G_q(t)$. The steps leading to (5.38) can no more be applied: <u>one never reaches a diffusion regime</u>.

More precisely, one can show that <u>$\Gamma'_q(t)$ shows an oscillatory approach to equilibrium with a characteristic time τ_q^{-1}</u>:

$$\tau_q^{-1} \propto q^{5/2} \tag{5.40}$$

While the oscillatory character of $\Gamma'_q(t)$ can only be obtained through detailed numerical analysis, the $q^{5/2}$ dependence is easily demonstrated by assuming $\Gamma_q(t) \equiv \Gamma(q^{5/2}t)$ and inserting this ansatz into Eqs. (5.22) and (5.36).

Let us point out that Eq. (5.40) is identical to the conclusion deduced at T_c from dynamical scaling (see (3.43)).

Similarly, when $T>T_c$, it can be shown that all predictions of dynamical scaling are exactly satisfied by the present model. The theory given here is however more precise than dynamical scaling because:

1) It furnishes an explicit determination of the function $\Omega(k/K_1)$ (see(3.20)), which, in dynamical scaling, is left unspecified.

2) It also offers a method for the explicit computation of the shape of the spectral function $f^A(x,k/K_1)$(see(3.21)), which, again, is not given by dynamical scaling.

However the analysis leading to these results involves heavy numerical calculation and is still the object of active investigation. We shall thus not discuss this here.

Finally, let us stress that, while our Weiss model is mathematically well defined, great care has to be taken when transposing its conclusion to realistic systems, where the number of neighbors is finite.

References: We limit ourselves to general reviews or to recent theoretical papers; from these, the reader can easily find the complete literature on the subject.

[1] M. FISHER: Rep. Prog. Phys. III, 615 (1967)
This brilliant review paper is an absolute prerequisite to any study of equilibrium critical phenomena.

[2] L. KADANOFF et.al.: Rev. Mod. Phys. 39, 395 (1967)
A detailed analysis of equilibrium scaling laws and their application.

[3] E. STANLEY: Introduction to Liquid-Gas and Magnetic Phase Transitions (to be published Oxford University Press, 1971)
A detailed and complete introduction to critical phenomena, requiring no background in the field.

[4] L. LANDAU AND E. LISHFITZ: Statistical Physics (London: Pergamon Press, 2nd. ed. 1969)
Its wonderful chapter on fluctuation theory remains the authoritative introduction to this field.

[5] L. VAN HOVE: Phys. Rev., 95, 1379 (1954)
This fifteen year old paper remains an example of how to apply simple quantum mechanics to complicated many body problems.

[6] W. MARSHALL AND R. LOWDE: Rev. Prog. Phys. XXXI, B, 706 (1968)
A review on magnetic correlation functions; although quite recent, this paper was written before most of the recent developments in critical phenomena; it is yet quite valuable as a lucid analysis of the assumptions involved in the classical theory.

[7] L. KADANOFF AND P. C. MARTIN: Ann. Physics, 29, 419 (1963)
A classical paper on the link between macroscopic theory and the modern correlation function approach.

[8] B. HALPERIN AND P. HOHENBERG: Phys. Rev. 177, 952 (1969)
A wonderful example of the far reaching consequences of simple but clever assumptions.

[9] L. KADANOFF AND J. SWIFT: Phys. Rev. 166, 89 (1968)
How to bypass the countless traps of N-body physics in order to solve a physical problem. See also the series of papers by K. KAWASAKI: Prog. Theor. Phys. Japan (1968).

[10] P. RESIBOIS AND M. DE LEENER: Phys. Rev. 152, 305, 318 (1966), 178, 806, 819 (1969)

See also K. KAWASAKI: J. Phys. Chem. Solids 28, 1277 (1968)
Prog. Theor. Phys. 39, 2, 285 (1968)

SOME EXACT RESULTS IN EQUILIBRIUM AND NON-EQUILIBRIUM STATISTICAL MECHANICS

J. L. Lebowitz
Belfer Graduate School of Science
Yeshiva University
New York, New York

CHAPTER I. INTRODUCTION

One of the developments in statistical mechanics in the last few years has been the study of rigorous results. The study of intensive properties of very large systems has been at the focal point of this subject because of the realization that many of the interesting phenomena peculiar to macroscopic systems, such as phase transitions and irreversibility, are intimately connected with and can be treated precisely only in the limit when the size of the system becomes infinitely large; called the bulk or the thermodynamic limit. Hence, (in studying these important phenomena), it is essential to discover whether the thermodynamic limit 'exists'. This question we shall discuss in lectures I and II. In the following lecture we shall discuss the problem of analyticity of the thermodynamic functions, and in the fourth lecture outline some results for non-equilibrium statistical mechanics.

I will only attempt here to sketch some of the problems, ideas and results in this area and refer you to "The Book" by Ruelle (1969) and also to the review article by Lebowitz (1968), for details and references. All parts of these lectures which parallel closely the discussion in my review article, as well as all references contained there, will be omitted from these notes.

CHAPTER II. EXISTENCE OF THE THERMODYNAMIC LIMIT

We start with a physical system and assume that its structure and properties can be described by a Hamiltonian. We will also assume that we can neglect nuclear forces and gravitational forces and still obtain a realistic picture of macroscopic matter under normal conditions. Nuclear forces are so strong and short range and hold the nuclei so tightly together that we do not expect ordinary matter to look any different if the nuclei were charged mass points. The gravitational forces on the other hand are so weak that the mutual gravitational interaction between particles in an ordinary sized object is negligible. As stated by Onsager (1967): thermodynamics is concerned with objects which are large compared to the size of a molecule, but small compared to the moon. The 'thermodynamic limit' should also be understood in this spirit. Ignoring also relativisitic effects, which we do not know how to take into account in any consistent way, we can write the Hamiltonian of a system of N particles of mass m as

$$H_N = \sum_{i=1}^{N} p_i^2/2m + V_N(\underline{r}_1,\ldots,\underline{r}_N) \qquad (2.1)$$

(This generalizes in an obvious way to a system of s species, of $N^{(j)}$ particles each, with masses m_j, j=1,...,s.)

Using the statistical mechanics of Gibbs, we define the canonical partition function of a system of N particles in a container Ω, of volume |Ω|, as

$$Z(\beta,N,\Omega) = \exp[-\beta A(\beta,N,\Omega)] = \exp[-\beta|\Omega|\ a\ (\beta,N/|\Omega|;\Omega)]$$

$$= \begin{cases} (N!)^{-1}(m/\beta h)^{3N} \int_\Omega\ldots\int d\underline{r}_1\ldots d\underline{r}_N \exp(-\beta V_N); & \text{classical system} \\ \text{trace} \exp[-\beta H_N] = \Sigma \exp[-\beta\ E\ (N,\Omega]; & \text{quantum system} \end{cases} \qquad (2.2)$$

where $\beta = 1/kT$, T is the temperature and $a(\beta,N/|\Omega|;\Omega) \equiv |\Omega|^{-1} A(\beta,N,\Omega)$ is the Helmholtz free energy per unit volume. We now want to determine whether this prescription for finding the thermodynamic free energy of a system from its microscopic Hamiltonian really leads to a proper thermodynamics for macroscopic systems. Thus, we ask the following questions. 1. Does the free energy density $a(\beta,N/|\Omega|;\Omega)$ as defined by eq. (2) have the property of not depending on the shape of the system in the "thermodynamic limit", that is when N and $|\Omega|$ tend to infinity and $N/|\Omega|$ tends to a definite density ρ?: i.e. given a sequence of containers Ω_j and particle numbers N_j, $|\Omega_j|\to\infty, N_j/|\Omega_j|\to\rho$ does $\lim a(\beta,N/|\Omega|;\Omega) = a(\beta,\rho)$ exist independently of the shape of the container Ω (as long as it is a 'reasonable shape'). 2. Assuming that $a(\beta,\rho)$ exists, is $a(\beta,\rho)$ a convex function of the density ρ and a concave function of the reciprocal temperature β? The last conditions will ensure the thermodynamic stability of our system. The answer to both these questions is yes.

The existence of the thermodynamic limit was proven some time ago for a large class of systems whose Hamiltonian satisfies two conditions. These two conditions are chosen so as to prevent the possibility of the system collapsing, as would happen in a gravitational system, or exploding like a system of positively charged particles would. The first condition is the 'H-stability' condition which requires that there be a lower bound on the energy per particle, i.e.

$$V_N(\underline{r}_1, \underline{r}_2, \ldots \underline{r}_N) \geq -NB;\ B < \infty, \text{ independent of } N \text{ and } \underline{r}_i \tag{2.3a}$$

When treating this system quantum mechanically, we would replace this condition by

$$E_0(N) \geq -NB \tag{2.3b}$$

where $E_0(N)$ is the ground state energy: (2.3a) implies (2.3b) but not conversely.

The second condition on the interaction potential is the tempering condition, and prevents the potential from being too positive at large separations and ensures against explosion. If we have two regions of space separated by a distance r, containing N_1 and N_2 particles respectively then the tempering condition requires that the interaction between the two groups have an upper bound of the form

$$V(N_1 \oplus N_2) - V(N_1) - V(N_2) \leq C\, N_1 N_2\, r^{-d-\varepsilon}, \text{ for } r > r_0 \qquad (2.4)$$

where C and r_0 are constants, d is the dimensionality of the system and ε is a positive constant. When V_N is a sum of pair potentials, $V_N(\underline{r}_1 \ldots r_N) = \Sigma\, v(\underline{r}_i - \underline{r}_j$ and v(r) is a Lennard-Jones type potential then both of these conditions are satisfied. We are thus all right for systems whose basic units are taken to be neutral 'spherical' atoms or molecules. This is not entirely satisfactory, however, in that we believe, as mentioned earlier, that the true interaction potential relevant for macroscopic matter is the Coulomb potential and we should be able to prove the existence of the thermodynamic limit directly for a system of Coulomb charges if the system is overall neutral (or 'approximately' neutral).

Now it is clear that a system of point charges whose pair interaction is $e_i e_j / |\underline{r}_i - \underline{r}_j|$ does not have a lower bound on its potential energy and hence does not satisfy (2.3a), (unless of course all the charges e_i are of the same sign which is not interesting since such a system clearly does not have any thermodynamics). When the particles have hard cores, however, i.e. there is a minimum distance of approach R, between the particles, then Onsager showed the existence of a lower bound of this type. (The following form of Onsager's proof was suggested to me by O. Penrose.) The Coulomb interaction energy between different particles can be written in the form

$$V_N = \frac{1}{2} \int \underline{E}^2 \, d\underline{r} - \sum_{i=1}^{N} U_i \geq - N \operatorname*{Max}_i U_i = -NB \qquad (2.5)$$

where $\underline{E}$ is the electric field and U_i is the self-energy of the i th particle which is finite if the charges are assumed to be distributed over a sphere of radius R. For real (point) charge particles however only the quantum version of H-stability, (2.3b) is possible and was recently proven by Dyson and Lenard. They show that (2.3b) will hold for any set of charges and masses provided that the negative particles and/or the positive ones are fermions. (It is curious that although stability of a small number of charged particles, say an atom, comes about mainly through the uncertainty principle, which keeps the oppositely charged particles apart, to obtain (2.3b) it is also necessary to keep the negative (or positive) particles apart from each other through the Pauli principle.)

The second requirement; that the potentials be 'tempered', is also obviously not satisfied by the Coulomb potential, which is a long-range potential. Thus, the proofs which make use of tempering do not apply. However, Lebowitz and Lieb (1969) have been able to overcome these difficulites and prove the existence of the thermodynamic limit also for Coulomb systems which are overall neutral.

We have also shown, when the system is not strictly neutral, that as long as the excess charge per unit surface area tends to zero as $|\Omega|\to\infty$, one always obtains the same canonical free energy as for the neutral system. If the excess charge per unit surface area tends to infinity, however, the free energy does not exist in the thermodynamic limit. And finally, if the excess charge per unit surface area tends to a constant then the free energy approaches a limit equal to the free energy of the neutral system plus the energy of a surface layer of the excess charge as given by elementary electrostatics.

The same methods which are used in the proof of the existence of the free energy density in the thermodynamic limit also show that this free energy density is thermodynamically stable. It is furthermore possible to show that the microcanonical and grand canonical ensembles yield thermodynamic potentials equivalent to those obtained from the canonical ensemble in the thermodynamic limit. Thus, it has been shown for a large class of systems, that the thermodynamic quantities as calculated from statistical mechanics are well defined in the bulk limit and have the required stability properties.

To obtain explicit expressions for these thermodynamic quantities from statistical mechanics it is generally necessary to resort to approximations or to formal power series expansions in the density or fugacity; the virial expansion. It has now been proven, (for systems with tempered potentials), that these expansions have a finite radius of convergence. Every system will thus be in a gas phase when its density is sufficiently low. At these low densities it is also possible to prove that the distribution functions have a convergent virial expansion.

CHAPTER III. PHASE TRANSITIONS

While the existence of the thermodynamic limit was initially proven for 'rigid wall' boundary conditions at the surface of Ω the results have recently been extended, for some systems, also to different boundary conditions; e.g. systems on a torus (periodic boundary conditions), and systems for which the normal derivative of the wave function vanishes on the surface of Ω (Fisher and Lebowitz, 1970; Robinson, 1970). There is little doubt that all free energy densities, (in the bulk limit), are independent of the boundary conditions. What is perhaps more interesting is that quantities like the magnetization per unit volume $m(\beta, h, \Omega)$, which is the <u>derivative</u> of the free energy density $a(\beta, h;\Omega)$ of a lattice spin system of volume $|\Omega|$ in the presence of an external magnetic field h , <u>do</u> depend sometimes on the boundary conditions even in the limit $|\Omega|\to\infty$. This was proven first by Peierls for a two dimensional Ising spin system with nearest neighbor ferromagnetic interactions at $h = 0$ and β large (low temperatures). The two different boundary conditions considered were (1) the spins at the surface all point up and (2) they all point down. Peierls' result implies the non-interchangeability of the order of taking the limit $|\Omega|\to\infty$ of $a(\beta, h;\Omega)\to a(\beta,h)$ and taking the derivative of $a(\beta, h;\Omega)$ with respect to h. It follows from this that the thermodynamic free energy density $a(\beta, h)$ will have, at low temperatures, a discontinuity in its first derivative, (the magnetization $m(\beta, h) = \partial a(\beta,h)/\partial h$), at $h = 0$, i.e. the system will have a phase transition. This can be seen simply (as pointed out to me by E. Lieb) from the fact that $a(\beta,h;\Omega)$ is, for different Ω, a sequence of convex functions of which approach a limit $a(\beta,h)$. The limit function must therefore also be convex and $\partial a(\beta,h;\Omega)/\partial h \to \partial a(\beta,h)/\partial h$ for all values of h at which the latter is continuous.

The Peierls argument and results have been extended to higher dimensions and more general kinds of spin and lattice gas systems (c.f. Ginibre 1970). This way the existence of 'phase transitions' in a variety of lattice systems has been proven. In addition, as is well known, the exact solutions of some two dimensional lattice models by Onsager and Lieb have explicitly shown the existence of phase transitions in these systems.

This is very satisfactory as far as it goes as it agrees with our experience of the ubiquity of phase transitions in macroscopic systems. What is less satisfactory, however, is the lack of proof, so far, of the existence of any phase transitions, such as the vapor-liquid transition, in continuum systems with reasonable pair potentials between the particles. It is only for the limiting case of an infinitely long range potential that the existence of a phase transition has been established rigorously in continuum systems.

I am referring here to the so-called van der Waals limit of a system whose pair potential has the form

$$v(r) = q(r) + \gamma^{d} \phi(\gamma r) \tag{3.1}$$

Here q(r) is a potential containing a hard core and satisfying the tempering condition and $\gamma^{d} \phi(\gamma r)$ is a Kac potential with range γ^{-1} such that

$$\int \gamma^{d} \phi(\gamma r) d\underline{r} = \alpha \tag{3.2}$$

independent of γ. It was shown by Kac, Uhlenbeck and Hemmer in one dimension, d=1, and later by Lebowitz and Penrose for any dimension that in the van der Waals limit $\gamma \to 0$, taken <u>after</u> the thermodynamic limit $|\Omega| \to \infty$ such a system will exhibit, for a large class of Kac potentials, a first order gas-liquid phase transition of the classical van der Waals type. The treatment of Lebowitz and Penrose has been extended recently to more general Kac potentials by Gates and Penrose (1969).

CHAPTER IV. NON-EQUILIBRIUM SYSTEMS

As can be seen from the earlier lectures, the rigorous study of equilibrium statistical mechanics has achieved notable results already. The comparable investigation of the infinite volume limit of non-equilibrium systems is much more difficult and has begun only recently. Results have been obtained by Lanford (1968a,b), but are restricted to one-dimensional systems of classical point particles interacting by smooth, finite range pair forces F.

Let (q_i, p_i) represent the positions and velocities of a set of particles of unit mass each. Then Newton's equations of motion have the form

$$\frac{d\, q_i(t)}{dt} = p_i(t), \qquad \frac{d\, p_i(t)}{dt} = \sum_{j \neq i} F(q_i(t) - q_j(t)) \tag{4.1}$$

where F is the interparticle force. If we have a finite number of particles then there is clearly a unique solution to this set of differential equations for all sets of initial conditions $\{q_i(0), p_i(0)\}$.

The existence of a meaningful solution to Newton's equations, i.e. the existence of a time evolution of the system, becomes however far from trivial when we consider a system consisting (in some limit) of an infinite number of particles. In such a system it is quite possible to begin with a perfectly reasonable set of initial values $\{q_i(0), p_i(0)\}$ and find after some finite time t that there are an infinite number of particles in a finite region of space and that the right side of (4.1) is infinite. We illustrate this with a simple example given by Lanford (1968a). If there are no interparticle forces and if at time zero, $p_i = -q_i$ for each i, then all the particles will be situated at the origin at time t=1. Thus, we need to find a class of initial conditions for which such catastrophies do not happen. In fact, as we are interested in equilibrium statistical mechanics, we would like to show that those classes of initial conditions which have non-zero probability of occurring in equilibrium, do not give rise to such catastrophies. An even stronger desired result is to show that the time evolution of a part of the system contained in a fixed region of space D will, at any time t, be determined entirely by the state of the system at time t=0 in the neighborhood of D (how large this neighborhood is will of course depend on t). This was indeed proven by Lanford for one dimensional systems. He proves the existence for all times of a "regular" solution of Newton's equations of motion for a "regular" initial configuration. A regular configuration is, roughly speaking, one in which the number of particles in a unit interval and the magnitude of the momentum of any particle in that interval have a bound of the form $\delta \log R$ where R denotes the distance of the interval from the origin. It is further shown that, at equilibrium, if either the activity is small or the interparticle potential is positive, the set of non-regular configuration has probability zero.

A question left open by these results is whether a state which at time t=0 is described by a set of correlation functions can still be described by a set of correlation functions when $t \neq 0$. This was investigated by Gallavotti, Lanford and Lebowitz (1970) who proved that, for certain classes of initial states, the time-evolving state is described by correlation functions and that these correlation functions satisfy the BBGKY hierarchy in the sense of distributions.

The initial states we consider can be described as follows: Suppose that the system is in equilibrium at temperature β^{-1} and activity z under the influence of a pair potential and an external potential h which is localized in a finite region I_h. At time t=0 we switch off the external field and the system begins to evolve. We prove that if the activity is sufficiently small (i.e. if we are deep inside the gaseous phase) the system can always be described by a set of correlation functions which vary in time according to the BBGKY hierarchy. We are, however, unable to prove even that the time averaged correlation functions evolve toward the correlation functions which correspond to the equilibrium state at temperature β^{-1} and activity z (in absence of external field) as would be expected. We are, however, able to prove that the time averaged correlation functions converge to a limit satisfying the stationary BBGKY hierarchy.

While initial states of the kind just described suffice, in principle, for the study of transport properties such as diffusion at low activity an alternative, sometimes more direct way to study transport processes is through the van Hove time displaced distribution functions (t.d.f.). These are time dependent correlation functions which correspond to different types of initial conditions from those just considered. Instead of considering the time evolution of an initial ensemble density having the form (for a fixed number of particles),

$$\mu(x_1,\ldots,x_N;\ t=0) = \mu_{eq}(x_1,\ldots,x_N)\ \Psi(x_1,\ldots,x_N)$$

where $x_i = (q_i,p_i)$, μ_{eq} is the equilibrium Gibbs canonical ensemble density corresponding to the correct Hamiltonian for $t > 0$ and Ψ is a symmetric function of the x_i; the t.d.f. are correlation functions obtained from an initial ensemble which is in equilibrium with one or more particles having specified positions and momenta. A typical ensemble density of this kind of

$$\mu(x_1,\ldots,x_N);\ t=0) = [\mu_{eq}(x_1,\ldots,x_N)/f_{eq}(x_1)]\,\delta(x_1-x_1^{0})$$

where $f_{eq}(x_1)$ is the equilibrium distribution function of particle one. The distribution function of this particle at time t, $f(x_1,t)$, is the time displaced one particle self distribution function. (If we integrate this function over velocities we obtain the van Hove self function which is important in neutron scattering experiments.) The self-diffusion constant can be obtained directly from $f(x,t)$, and other transport coefficients can be obtained from similarly defined t.d.f.

We (Lebowitz and Percus, 1967: Lebowitz, Percus and Sykes, 1968, 1969) have made an extensive study of these t.d.f. for a one dimensional system of hard rods of diameter R obtaining many of them in explicit form. The self-diffusion constant for example, (defined, of course, in the thermodynamic limit), is given by $D = (2\pi\beta m)^{-1/2}(1-\rho R)/\rho$. Much remains to be done in this field.

ACKNOWLEDGMENTS

This research was supported by the U.S.A.F.O.S.R. under Grant No. 68-1416 and Contract F44620-71-C-0013. Notes at the lectures were taken by R. W. Gibberd and R. H. G. Helleman. I would also like to thank Professors I. Prigogine and William C. Schieve for their kind hospitality and stimulating discussions.

References:

[1] M. E. FISHER AND J. L. LEBOWITZ: "Asymptotic Free Energy of a System With Periodic Boundary Conditions", to appear in Communications in Mathematical Physics, 1970.

[2] G. GALLAVOTTI, O. E. LANFORD III, AND J. L. LEBOWITZ: "Thermodynamic Limit of Time Dependent Correlation Functions for One Dimensional Systems", to appear in Journal of Mathematical Physics, 1970.

[3] D. J. GATES AND O. PENROSE: Communications in Mathematical Physics 15, 255 (1969): 16, 231 (1970).

[4] J. GINIBRE: Colloques Internationaux Du Centre National De La Recherche Scientifique 181, 163 (1970).

[5] O. E. LANFORD (1968): Communications in Mathematical Physics 9, 176 (1968): 11, 257 (1969).

[6] J. L. LEBOWITZ: Annual Review of Physical Chemistry 19, 389 (1968).

[7] J. L. LEBOWITZ AND E. H. LIEB: Physical Review Letters 22 631 (1969).

[8] J. L. LEBOWITZ AND J. K. PERCUS: Physical Review 155, 122 (1967).

[9] J. L. LEBOWITZ, J. K. PERCUS AND J. SYKES: Physical Review 171, 224 (1968): 188, 487 (1969).

[10] L. ONSAGER: The Neurosciences, Rockefeller University Press (1967).

[11] D. ROBINSON: Communications in Mathematical Physics 16, 290 (1970).

[12] D. RUELLE: Statistical Mechanics (Benjamin, New York, 1968)

INTRODUCTION TO NON-EQUILIBRIUM STATISTICAL MECHANICS

Radu Balescu
Universite Libre de Bruxelles
Belgique

CHAPTER I. THE LIOUVILLE EQUATION

The purpose of statistical mechanics, as everybody knows, is the description of the mechanics of large systems, or rather, large assemblies of microscopic systems, such as molecules. Of course, because of the number of component systems being so large, we cannot hope to have a description which would be exact in the same sense as for two-body systems, say, or even for the systems of celestial mechanics in which there could be 10 or 12 bodies. But here in statistical mechanics, the purpose (and therefore the methods) are essentially different, because we are interested in this different type of system. We could say that one of the aims of statistical mechanics is to devise a method by which the exact description of the system is progressively contracted. So one gets through various stages in the development and at each stage some information is lost. We go from the exact description to a less exact description which is however sufficient for the purposes one is interested in; then maybe at another stage one can throw away some more information and have a more contracted description. By such successive contractions one can find formulas which are more and more useful for specific purposes. The only thing one has to worry about is to make these contractions in a clever way.

Let me summarize the steps which appear in the next lectures. We start with an essentially exact description of the system, the Liouville equation, and we write down its formal solution. The advantage of doing so is that the Liouville equation is linear. And of course one has methods in mathematics to deal with linear equations in an exact way. However this formal solution will not help us very much because we are just not interested in an exact description of the system. Even if we could handle the solution, it would be in terms of an initial condition which we could never measure. So this is not the point of interest to us. We will then study how this solution behaves under certain limiting conditions. In particular, we are considering large systems: this circumstance will allow us to simplify, to throw away some part of the solution which is not interesting to us. In the same sense, we will be interested in times which are very long compared to some elementary dynamical times in the system. Again, by studying the behavior of the solutions under this limiting condition we will be able to throw away part of the information which is only interesting for very short times and therefore is not accessible to our experience. By throwing away these terms, we can get a more contracted form of the solution. So this will be essentially our problem.

Let me summarize first some of the basic formulas of statistical mechanics. Statistical mechanics essentially links together two descriptions of nature. One is a description at the microscopic level. Its basic building stones are the dynamical functions $a(p,q;\underset{\sim}{x})$ defined in a phase space $(p_1 \ldots p_n,\ q_1 \ldots q_n) \equiv (p,q)$ characteristic of the system. These functions may also depend on a position coordinate $\underset{\sim}{x}$ which is important when we consider spatially inhomogeneous problems. As an example, the microscopic density function is

$$\rho(q,p;\underset{\sim}{x}) = \sum_i \delta(q_i - \underset{\sim}{x}),$$

where q_i are the phase space coordinates of the particles in the system.

Another description is the macroscopic one. Here the relevant functions are observable quantities, defined in the physical, four dimensional space: $\underset{\sim}{x}$, t.

Statistical mechanics provides a link between the two descriptions by introducing the concept of a <u>distribution function</u> $f(pq;t)$. It is then postulated that the following relation connects microscopic quantities $a(pq;\underset{\sim}{x})$ and corresponding macroscopic quantities $A(\underset{\sim}{x};t)$:

$$A(\underset{\sim}{x};t) = \int dpdq \;\; a(pq;\underset{\sim}{x}) f(pq;t) , \tag{1.1}$$

the integration being over the whole accessible phase space. There is a consistency condition which we could require from the formula (1.1). The average of a constant must be a constant. And therefore this implies

$$\int dpdq \; f(p,q;t) = 1 , \tag{1.2}$$

at all times. We may also require the distribution function to be semi-definite positive

$$f(p,q;t) \geq 0 . \tag{1.3}$$

In this case $f(pq;t)$ can be interpreted as a probability density distribution. Now this is usually assumed in classical statistical mechanics but one can show that perfectly coherent formulas can be constructed in which condition (1.2) is satisfied but not (1.3). A typical example is the description of quantum statistical mechanics in terms of Wigner functions. So ineq. (1.3) will not play an important role in the theory presented here.

Let us now introduce the dynamics. The distribution function $f(pq;t)$ obeys the fundamental Liouville equation

$$\frac{\partial f}{\partial t} = [H,f]_p \equiv Lf , \tag{1.4}$$

where $[,]_p$ is the Poisson bracket, and H is the hamiltonian of the system. We may now calculate the rate of change in time of an average quantity:

$$\frac{\partial A(\underset{\sim}{x};t)}{\partial t} = \int dpdq\ a(pq;\underset{\sim}{x})\ \frac{\partial f(pq;t)}{\partial t} = \int dpdq\ a(pq;\underset{\sim}{x})\ Lf(pq;t). \qquad (1.5)$$

We may then perform an integration by parts in the right hand side (L is a first-order differential operator) and obtain

$$\frac{\partial A(\underset{\sim}{x};t)}{\partial t} = -\int dpdq\ f(pq;t)\ \{La(pq;\underset{\sim}{x})\} \equiv B(\underset{\sim}{x};t). \qquad (1.6)$$

Indeed, La is some new microscopic dynamical function, and therefore we have expressed $\partial A/\partial t$ in terms of a new macroscopic function B. We may now generate a hierarchy of equations, by expressing $\partial B/\partial t$ in terms of some C, and so on. If we could solve this hierarchy, we could say that the final contraction has been achieved. This however is an illusion.

We can easily see that by proceeding in a very direct way, as we did, we have not solved the problem at all. It is true that the equations are entirely in a four-dimensional physical space, however there is an infinity of them. So we have not achieved any simplification at all.

From this point on then, we can proceed in several ways. One way is to assume that we may cut the hierarchy at some stage by making assumptions about approximate functional dependencies of the averages B,C in terms of averages that have been calculated at the lower stage (say, A). If we can close the hierarchy in such a way, then of course we are left with a finite number of equations which are contracted and we can go ahead. This is precisely what is done in hydrodynamics. In hydrodynamics when one writes equations for successive moments of the distributions one also gets hierarchies, for instance density is related to local velocity and the local velocity will be related to the second moment, etc. One then cuts the hierarchy by expressing the second moment in terms of the first one. This is an intuitive physical argument, which is not necessarily justified ; one of the purposes of statistical mechanics is to justify such types of closures. A rather similar type of method appears in non-equilibrium statistical mechanics in the study of the BBGKY hierarchy. The BBGKY hierarchy is essentially equivalent to the Liouville equation. One can define successive reduced distribution functions,and writing the equation for a reduced one-particle distribution function one should know the two-particle distribution. So we generate a hierarchy,and by making some guesses we can approximately cut the hierarchy into various approximations. Now the method that I expose in these lectures is different. The starting point of the method is to exploit essentially the linearity of the Liouville equation. For a linear equation we can write a formal exact solution. And next we will try to examine its asymptotic behavior, i.e. to examine the behavior of the solution of the Liouville equation under certain limiting conditions such as large systems, long times, and so on. Hence, we try first to simplify (or to contract) the Liouville equation description and to get some approximate descriptions valid under asymptotic conditions, and then to use the solution of the simplified equations of evolution to calculate averages like (1.1) and (1.6).

Before going on with details let me just mention very briefly how the problem is seen in quantum mechanics (because all of this was classical mechanics). In quantum mechanics, in the most familiar form, the distribution function f is essentially replaced by von Neumann's density matrix $\underline{\underline{\rho}}$. The prescription for taking averages of dynamical operators $\underline{\underline{a}}$, replacing (1.1), is

$$A(\underline{x},t) = \mathrm{Tr}\ \underline{\underline{\rho}}(t)\underline{\underline{a}} \quad , \tag{1.7}$$

with the normalization condition

$$\mathrm{Tr}\ \underline{\underline{\rho}}(t) = 1 \quad . \tag{1.8}$$

The dynamical equation replacing the Liouville equation (1.4) is

$$i\hbar \frac{\partial \underline{\underline{\rho}}}{\partial t} = [\underline{\underline{H}}, \underline{\underline{\rho}}] \ , \qquad (1.9)$$

where [,] is the commutator. From here on, everything is quite parallel to the classical case. Now let us come back to the Liouville equation, and let me make certain remarks about the Hamiltonians. In many problems there exists a natural decomposition of the Hamiltonian into two terms:

$$H = H^{o} + \lambda H' . \qquad (1.10)$$

If you take systems of particles then the decomposition corresponds to the decomposition into kinetic energy and interaction energy

$$H^{o} = \sum_{i} \frac{p_i^2}{2m} \qquad (1.11)$$

$$H' = \lambda \sum_{i<j}\sum V(q_i - q_j) \ ,$$

where we assume that the intermolecular interactions derive from a two-body potential function λV, depending only on the relative distance between the particles. λ is a dimensionless parameter measuring the size of the perturbation. Now let us stress immediately the main feature here: H^{o} is a <u>sum of independent terms</u>, each describing a single degree of freedom. In parallel with (1.10) we can write a quite similar decomposition of the Liouville operator

$$L = L^{o} + \lambda L' \ , \qquad (1.12)$$

where L^{o} derives from H^{o} and L' from H' by eq. (1.4). Again, L^{o} is a <u>sum</u> of operators involving a single degree of freedom, whereas L' does not have this property.

The simplest method to solve the Liouville equation is to use a Laplace transform technique. We write the distribution function in the form

$$f(t) = \frac{1}{2\pi} \int_C dz\, e^{-izt}\, \tilde{f}(z)\,, \tag{1.13}$$

where C denotes a contour parallel to the real axis and lying above all the singularities of $f(z)$. We then introduce the resolvent operator $R(z)$ of the Liouville equation which, by definition, connects the Laplace transform $\tilde{f}(z)$ to the initial value $f(0)$:

$$f(t) = \frac{1}{2\pi} \int_C dz\, e^{-izt}\, R(z)\, f(0)\,. \tag{1.14}$$

The resolvent can be written formally as

$$R(z) = (-L-iz)^{-1}. \tag{1.15}$$

Using the decomposition (1.12) we can expand the resolvent in the form:

$$R(z) = \sum_{n=o}^{\infty} \lambda^n R^o(z)\, [L' R^o(z)]^n\,, \tag{1.16}$$

where $R^o(z)$ is clearly defined as : $R^o(z) = (-L^o-iz)^{-1}$ (1.15a). Equation (1.16), substituted into (1.14), provides the formal solution of the initial value problem for the Liouville equation in the form of an infinite perturbation series:

$$f(t) = \sum_{n=o}^{\infty} \lambda^n \frac{1}{2\pi} \int_C dz\, e^{-izt}\, R^o(z)\, [L' R^o(z)]^n\, f(0)\,. \tag{1.17}$$

This equation is a basic step in our theory. Let me say at once that I will not discuss the convergence properties of such things. On the other hand one of the main features of statistical mechanics is the following: In elementary quantum theory one essentially makes developments of this sort and then one stops after one or two orders of magnitude and one can derive in some cases good results. In statistical mechanics, and in general in the many-body problem, perturbation expansions which are limited to a few terms usually are insignificant. One can only do perturbation theory by going to all orders. Only in that case can one show that the result becomes significant. This is essentially because even if the parameter λ is small, each of these terms involves sums over all degrees of freedom, i.e. sums of a very large number of terms. Therefore, for a large system, even if individual terms in (1.17) may be small it is by no means clear that the global coefficient of a given order in λ is indeed small. We must therefore proceed to a much more careful analysis.

We have first to make our method operational, and therefore give a meaning to the various objects appearing in (1.17). To do so, we first choose a natural representation (as one does in quantum mechanics). Consider again the case of a classical gas. It is easily seen that if we represent the distribution function as a multiple Fourier series with respect to the q's (assuming periodic boundary conditions):

$$f(qp;t) = \sum_k e^{ikq} \rho_k(p;t) , \tag{1.18}$$

then the unperturbed Liouville equation

$$\frac{\partial f}{\partial t} = L^o f \equiv -v\frac{\partial}{\partial q} f \qquad (v \equiv \rho/m) ,$$

goes over into

$$\frac{\partial \rho_k}{\partial t} = -ikv\rho_k \quad . \tag{1.19}$$

Hence the unperturbed Liouvillian L^o reduces to a purely algebraic operator in this representation. We therefore obtain immediately a representation of the matrix element of R^o:

$$\langle k|\overset{o}{R}(z)|k'\rangle = \frac{1}{ikv-iz}\,\delta_{k-k'} \, . \tag{1.20}$$

In all these formulae, we use abbreviations: $k \equiv (\underset{\sim}{k}_1,\ldots,\underset{\sim}{k}_n)$; $kv \equiv \sum_{i=1}^{N} \underset{\sim}{k}_i \cdot \underset{\sim}{v}_i$, etc. We can now calculate matrix elements of the perturbation:

$$\langle k|L'|k'\rangle \equiv \int dq\, e^{-ikq} L'\, e^{ik'q}$$

$$= \sum_{j<n} \frac{8\pi^3}{\Omega} V_{|\underset{\sim}{k}-\underset{\sim}{k}'|} \frac{i}{m}(\underset{\sim}{k}_j - \underset{\sim}{k}'_j)\cdot\left(\frac{\partial}{\partial \rho_j} - \frac{\partial}{\partial \rho_n}\right) \delta_{\underset{\sim}{k}_j+\underset{\sim}{k}_n-\underset{\sim}{k}'_j-\underset{\sim}{k}'_n} \prod_{n\neq j,n} \delta_{\underset{\sim}{k}_m-\underset{\sim}{k}'_m}, \qquad (1.21)$$

where V_k is the Fourier transform of the potential $V(\underset{\sim}{q}_j-\underset{\sim}{q}_n)$; Ω is the volume. Let us note for later use, that summations over k go over values

$$\underset{\sim}{k} = \frac{2\pi}{\Omega^{1/3}}(n_1,n_2,n_3) \quad ,$$

where n_1's are positive or negative integers. At the end of the calculations we shall let $\Omega\to\infty$; the allowed values of $\underset{\sim}{k}$ then become dense, and we may then use the correspondence:

$$\sum_{\underset{\sim}{k}} \to \frac{\Omega}{8\pi^3} \int d\underset{\sim}{k} \quad . \qquad (1.22)$$

We can now write explicitly eq. (1.17) in the Fourier representation:

$$\rho_k(t) = \sum_{n=0}^{\infty} \lambda^n \frac{1}{2\pi} \int_c dz\, e^{-izt} \sum_{k'} \langle k|R^0(z)\{L'R^0(z)\}^n|k'\rangle\, \rho_{k'}(0) \quad . \qquad (1.23)$$

Eqs. (1.20) and (1.21) provide us with everything we need for writing down every term of this series.

Let me now make a few remarks on the translation of these results in quantum mechanics. We start from the von Neumann equation (1.9) written in the occupation number representation (second quantization):

$$ih\frac{\partial}{\partial t} \langle n|\rho|n'\rangle = \sum_{n''} \{\langle n|H|n''\rangle\langle n''|\rho|n'\rangle - \langle n|\rho|n''\rangle\langle n''|H|n'\rangle\} \,, \qquad (1.24)$$

and introduce new variables:

$$n - n' = \nu; \quad n + n' = N \quad .$$

We then write the matrix elements of an arbitrary operator $\underset{=}{A}$ as follows:

$$\langle n|\underline{A}|n'\rangle \equiv A_{n-n'}(n+n') \equiv A_\nu(N) \quad .$$

It is easily seen that (1.24) can be written as

$$\partial_t\rho_\nu(N) = \sum_{\nu''}\{H_{\nu-\nu''}(N+\nu'')\rho_{\nu''}(N+\nu''-\nu) - H_{\nu-\nu''}(N-\nu'')\rho_{\nu''}(N+\nu-\nu'')$$

or else, introducing finite displacement operators:

$$\partial_t\rho_\nu(N) = \sum_{\nu'}\{e^{\nu'\frac{\partial}{\partial N}} H_{\nu-\nu'}(N)e^{-\nu\frac{\partial}{\partial N}} - e^{-\nu'\frac{\partial}{\partial N}} H_{\nu-\nu'}(N)e^{\nu\frac{\partial}{\partial N}}\}\rho_{\nu'}(N) \; .$$

We now <u>define</u> a quantum Liouville operator by identifying its matrix elements with the bracketed expression:

$$\partial_t\rho_\nu(N) \equiv \sum_{\nu'}\langle\nu|L|\nu'\rangle\rho_{\nu'}(N) \quad . \qquad (1.25)$$

In this way von Neumann's equation is written in the same form as the classical Liouville equation in Fourier representation, and the whole formalism can be easily adapted to this case.

CHAPTER II. VACUUM AND CORRELATIONS

In order to do explicit calculations, I introduced yesterday a Fourier representation and in terms of that representation the matrix elements of the resolvent appear in equation (1.23). Here we have of course matrix elements of a product. Therefore if we write them in detail for some order of λ we will have transitions of the type

$$\langle k|R^{o}|k\rangle\langle k|L'|k^{1}\rangle\langle k^{1}|R^{o}|k^{1}\rangle\langle k^{1}|L'|k^{2}\rangle \cdots \langle k^{n}|L'|k'\rangle\langle k'|R^{o}|k'\rangle.$$

Now the intermediate states $(k^{1},k^{2},\ldots)$ effectively represent Fourier components of the distribution function,and these Fourier components have in general some simple physical interpretation in terms of correlations between the various degrees of freedom . This can be seen rather easily.

If we classify the Fourier components according to the number of non-vanishing wave-vectors they involve , we could write the decomposition of f in the form:

$$f(q,p) = \rho_{o}(p)+\sum_{j}\sum_{\underset{\sim}{k}}{}' \rho_{k}(p_{j}|p\ldots)e^{ik\cdot q_{j}}+\sum_{j<n}\sum_{\underset{\sim}{k}}{}'\sum_{\underset{\sim}{k}'}{}' \rho_{kk'}(p_{j}p_{n}|p)e^{ik\cdot q_{j}+ik\cdot q_{n}}. + \ldots\ldots \qquad (2.1)$$

The primes over the k-summation signs mean that the value $k = 0$ is excluded. For brevity, we have omitted writing the necessary volume factors in this formula, because they are irrelevant in our brief discussions.

It is clear that $\rho_{o}(p)$ is simply the integral of $f(q,p)$ over all positions; in other words, it is the reduced momentum distribution function. The second term, involving one wave-vector, represents density fluctuations: we can call it an inhomogeneity factor. It is essentially a measure of the degree of inhomogeneity of the system, because it depends on a single position and is not translationally invariant. Therefore if the system is inhomogeneous, then necessarily

$\rho_{k} \neq 0$.

Now when we come to components with more than two wave vectors, these will clearly describe correlations between degrees of freedom. For instance $\rho_{kk'}$, involving two wave-vectors will describe two-body correlations; the components with more than two wave-vectors will describe more complicated patterns of correlations between the particles. In the case of homogeneous systems, by translational invariance, we see that the only non-vanishing $\rho_{kk'}$ component is the one for which $k = -k'$. So in that case the Fourier component $\rho_{k,-k}$ represents pure correlations. In the case of inhomogeneous systems however, the interpretation in terms of correlations is not quite correct, because there is also a term which involves just a product of inhomogeneity factors:

$$\rho_{kk'} = \rho_{kk'}^{(c)} + \rho_{k}\rho_{k'} \quad . \tag{2.2}$$

So we see there is a mixture of uncorrelated and correlated parts. Similar considerations apply to all other Fourier components.

My purpose was to show that in this way of describing the dynamics, based on eq. (1.23), the picture is one of transitions from one state of correlation to another. And of course as a particular case, there are states without correlations at all, for instance: ρ_0, ρ_k. These we call vacuum states.

What I would like to do today is to make these ideas a little more precise and more formal. I shall start with some quite general remarks about the structure of distribution functions. First of all, there are some introductory properties which may seem rather trivial, but it is not bad to keep them in mind in order to understand the structure of the theory, and of statistical mechanics in general.

We introduced the concept of distribution function in eq. (1.1). Let us call F the set of all phase-space functions which can play the role of distribution functions. The only condition we shall require of these functions is that they be normalized to one:

$$\int d\Gamma\, f = 1 : \quad f \in F , \qquad (2.3)$$

where $d\Gamma \equiv dpdq$.

Let us now make the idea of independent degrees of freedom more precise. We should note that we have already met previously the concept of independence of degrees of freedom. This occurred when we introduced the separation of the Hamiltonian H into $H^0 + \lambda H'$: eq.(1.11) We have defined H^0 as a sum of terms, each involving a single degree of freedom. Similarly:

$$L^0 = \Sigma L^{0(j)} .$$

If we want to define a decomposition of the distribution functions into vacuum and correlations, this decomposition should clearly be connected in some way with the dynamical concept of independence. So let us see how we can do this in a general way.

Our purpose is to separate f into two parts which we write as:

$$f = Vf + Cf , \qquad (2.4)$$

where V and C are operators to be presently defined. In order to simplify the presentation we consider only homogeneous systems.

Let us now assume that a system is described at time zero by a pure vacuum distribution function, $f = Vf$. If this state moves according to L^0 alone, then at any later time, there can be no correlations in the system. In other words, L^0 cannot create correlations out of a vacuum. The only physical mechanism that can create correlations is the interactions. So this appears to be a very convenient way of defining correlations in an axiomatic way. We express this idea by requiring that the unperturbed propagator $\exp(L^0 t)$ commutes with the vacuum operator V:

$$e^{L^o t} V f = V e^{L^o t} f, \; f \in F, \text{ all } t \; . \tag{2.5}$$

We may also write (2.4) as an operator equation:

$$V + C = I \; . \tag{2.6}$$

We demand also the decomposition (2.6) to be unique, i.e. that V and C are mutually exclusive. This is ensured by the axiom

$$V^2 = V \; , \tag{2.7}$$

from which, together with (2.6) follows:

$$C^2 = C \qquad VC = 0 \qquad CV = 0 \; . \tag{2.7a}$$

Because V and C have all the properties of familiar <u>projection operators</u>, eqs. (2.4) - (2.7) will be taken as our basic set of axioms defining the projection operators. We will see that these properties are sufficient for the derivation of the master equation.

We readily verify that

$$V \partial_t f(t) = \partial_t V f(t) \; . \tag{2.8}$$

This commutation relation turns out to be quite important in subsequent developments.

In the case of homogeneous systems one would like to identify in some way the vacuum with the Fourier component ρ_0, because we know intuitively that all other Fourier components represent pure correlations. We therefore try the definition:

$$V f = \rho_0 \; . \tag{2.9}$$

It is an easy matter to verify that all the axioms (2.4-2.7) are satisfied by this choice. So this is a good realization of the vacuum and it corresponds to physical intuition. Another important feature is that (2.9) is a <u>linear</u> realization: $Vf = \int dq \; f(qp)$.

CHAPTER III. THE MASTER EQUATION

Let me go back now to the formula which we have derived the first day; i.e. the expression of the resolvent of the Liouville equation as a power series in λ, eq. (1.17). I said before that one can calculate out these expressions by going over to a Fourier representation; but this is not really necessary. Actually, by using our abstract projection operators we can go through the calculations in a quite general way, independent of any particular representation. Only for explicit calculations, at the end of the formal derivations, should we choose a specific representation such as (2.9) or any other convenient representation. One of the advantages of the projection operators is precisely to allow us to do the calculations independently of any particular representation. And this leads to more compactness and perhaps more transparency.

What we shall do is to rewrite the series (1.17) in a different form, more relevant to our purpose. The crucial idea here is the separation into vacuum and correlations. Now technically the developments are very simple. We first note the characteristic structure of (1.17) in which one has an alternation of factors R^0 and L .

$$R^0 L' R^0 L' R^0 \ldots L' R^0 .$$

We shall first rewrite the series by inserting between each two successive factors the unity in the form $I = V + C$ (see 2.6) for instance:

$$R^0 L' R^0 L' R^0 = R^0 (C+V) L' (C+V) R^0 (C+V) L' (C+V) R^0 \quad . \tag{3.1}$$

And now we can rearrange the whole series. We can write the single product which appears in (3.1) as the sum of many products each of them being of the typical form:

$$R^oCL'VR^oVL'CR^o \quad . \tag{3.2}$$

So this is a typical term and we have all possible distributions of V and C between R^o and L' factors. Certain terms in (3.1) will not contribute at all. For instance:

$$R^oCL'VR^oCL'CR^o \quad .$$

Indeed, this term contains the operator VR^oC. It is easily seen that as a result of the commutation relation (2.5)

$$VR^oC \equiv 0 \quad .$$

We can now rearrange the whole sum by partitioning every product of type (3.2) at the points where a factor V appears, and grouping together all possible products occurring between successive V-factors; we then sum over all possible numbers of V-factors:

$$R(z) = \sum_{p=0}^{\infty} R^o(z)[E(z)\, VR^o(z)]^p \, [1 + E(z)\, CR^o(z)] \, , \tag{3.3}$$

where E(z), the "irreducible evolution operator", is defined as

$$E(z) = \sum_{m=0}^{\infty} \lambda^m L' [CR^o(z)L']^m \quad . \tag{3.4}$$

We now have a description which is really in terms of "dynamics of correlations." Let us study the vacuum part of f: We then project eq. (1.14) into the vacuum and use (3.3) :

$$Vf(t) = \frac{1}{2\pi} \int_C dz\, e^{-izt} \sum_{p=0}^{\infty} VR^o[EVR^o]^p[V + VER^oC]f(0) \, . \tag{3.5}$$

We start with a vacuum on the left and then we have a number of transitions through correlations then back to the vacuum; and then again to correlations and back to the vacuum, etc. So we have decomposed the whole resolvent into irreducible transitions from vacuum to vacuum. And then at the last step either one ends on the vacuum or there is another irreducible transition which brings us to a correlation state. So this is the general structure.

And now we have practically derived the master equation. Indeed, using (1.13)(1.14) and (3.3) we can easily derive the recurrence relation:

$$\tilde{f}(z) = R^{0}(z)E(z)V\tilde{f}(z) + R^{0}(z)E(z)R^{0}(z)Cf(0) + R^{0}(z)f(0). \qquad (3.6)$$

On substitution into (1.13), we easily get by time differentiation:

$$\partial_t f(t) = L^{0}f(t) + \frac{1}{2\pi}\int_C dz\, e^{-izt}\{E(z)Vf(z) + E(z)R^{0}(z)Cf(0)\} \qquad (3.7)$$

And finally, by inverting the Laplace transformation and using:

$$U^{0}(t) = e^{tL^{0}} = \frac{1}{2\pi}\int_C dz\, e^{-izt}R^{0}(z) \qquad (3.8)$$

$$E(t) = \frac{1}{2\pi}\int_C dz\, e^{-izt}E(z) \quad , \qquad (3.9)$$

we obtain the <u>master equation</u>, first derived by Prigogine and Resibois in 1961:

$$\partial_t f(t) = L^{0}f(t) + \int_0^t d\tau\, E(\tau)Vf(t-\tau) + \int_0^t d\tau\, E(\tau)U^{0}(t-\tau)Cf(0). \qquad (3.10)$$

We note that no approximation has been made in the derivation: Eq. (3.10) is just a sophisticated way of writing the Liouville equation. It makes however quite explicit the separate behavior of the vacuum and of the correlations. A close analysis of the derivation indeed shows that all the axioms listed before are necessary for the proof of eq. (3.10). Let us now examine more closely the dissymmetry between vacuum and correlations, which plays a crucial role in the theory. If we project eq. (3.10) on the vacuum we get:

$$\partial_t Vf(t) = L^{0}Vf(t) + \int_0^t d\tau\, VE(\tau)Vf(t-\tau) + \int_0^t d\tau\, VE(\tau)U^{0}(t-\tau)Cf(0). \qquad (3.11)$$

This is a closed integro-differential equation for the component $Vf(t)$. The kernel operator $VE(\tau)V$ is called the diagonal fragment. We note the presence of a source term, called the destruction fragment: it is a functional of the initial value of the correlations (which is of course a known function). So this is the first very important feature: the vacuum part of the distribution function obeys a closed equation.On the contrary (and this is another interesting part) you see that the correlation component does not obey a closed equation. By projecting (3.10) on the correlations, we see that $\partial_t Cf(t)$ is expressed in terms of $Vf(t)$ and of $Cf(t)$.

Let us remark that the problem is well-posed mathematically. We have first a closed equation for the vacuum, we have to solve that equation, and having solved it we can put the vacuum in the correlation equation and solve that equation.

Now let's see some other features of this equation. A very important feature is the fact that (3.11) is a non Markoffian equation. In other words the rate of change of Vf depends on the past values of the same function. So there is a memory in the system and the duration of the memory is essentially determined by the function $VE(\tau)V$.

Let us now examine the asymptotic behavior of this equation for long times. We shall do this very quickly here, and take up the problem in more detail in the next lectures. In the simplest cases it can be shown that the memory kernel $VE(\tau)V$ has a finite extension in time (see Figure 1). The effective length of the memory can be correlated in these cases to the duration of a collision, t_c. Hence, we may assume: $VE(\tau)V \to 0 \quad \text{for} \quad \tau >> \tau_c \quad .$ (3.12)

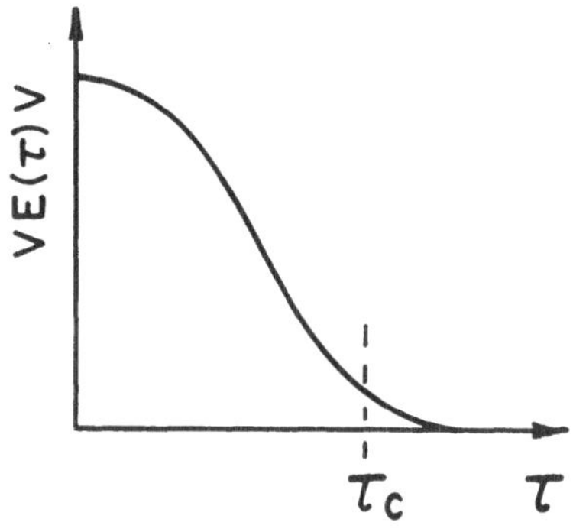

FIGURE 1

On the other hand, if the initial correlations are sufficiently well behaved - and we shall discuss later what this means - it can be shown that (3.12) implies a similar property for the destruction fragment in eq. (3.11):

$$D(t) \equiv \int_0^t dt\ VE(\tau)CU^{o}(t-\tau)f(0) \to 0,\ t >> \tau_c\ . \qquad (3.13)$$

It then follows that, for times,much longer than the duration of a collision, that the vacuum part of the distribution function $\bar{f}(t)$ obeys the simpler equation.

$$\partial_t V\bar{f}(t) = L^{o}V\bar{f}(t) + \int_0^\infty d\tau VE(\tau)V\bar{f}(t-\tau)\ . \qquad (3.14)$$

Equation (3.14) has been derived from eq. (3.11) by making use of the assumptions(3.12) and (3.13). It will be called the <u>general kinetic</u> equation, and we will see that it has properties quite different from the original master equation. In particular, we will see that irreversibility has entered the theory precisely at this point.

CHAPTER IV. THE WEAKLY COUPLED GAS

My purpose today is to illustrate in the case of a very simple example the rather abstract concepts which we have discussed yesterday. In this simple case, one can go into much detail and see under what conditions all those concepts can be valid and in particular what the ingredients of the passage from reversible to irreversible equations are. So the matter I will discuss today is the case of a classical gas, which is sort of an artificial gas, because we will assume that it is <u>weakly coupled</u>. In other words, the parameter λ is assumed small. Now this is rather unrealistic because whenever the particles come very close together in a collision their interaction is very strong. There exist however systems which approach the condition of weak coupling. This is the case of plasmas. Indeed, when the forces have very long range, two particles never have a chance (on the average) of coming very close together because they are deflected when they are still at a large distance from each other; on the other hand at large distances, the interactions are indeed small. It is true on the other hand that the long range of the Coulomb forces introduces other difficulties, but we shall not go into these problems here.

Not only shall I assume that the gas is weakly coupled, which is a dynamical assumption, but I shall also make a statistical assumption by considering a homogeneous system. So, our model will be a <u>homogeneous weakly coupled gas</u>.

We will now write explicitly the master equation in this approximation. First of all, we choose a realization of the projection operators; this will clearly be eq. (2.9). We write out explicitly eq. (3.10) to the lowest order in λ. We note that

$$VL^{o}Vf = <0|L^{o}|0>\rho_{o} \equiv 0$$
$$VL'Vf = <0|L'|0>\rho_{o} \equiv 0 \quad . \tag{4.1}$$

These properties follow from eqs. (1.19) and (1.21). Hence the lowest order non trivial term for VEV is in λ^2 and is given by:

$$VE(\tau)V = \frac{\lambda^2}{2\pi} \int dz\ e^{-izt} \sum_{i<j} \frac{8\pi^3}{\Omega} \sum_{k} <0|L'_{ij}|\underset{\sim}{k},-\underset{\sim}{k}> \frac{1}{ik(v_i-v_j)-iz} <\underset{\sim}{k},-\underset{\sim}{k}|L'_{ij}|0> . \tag{4.2}$$

We now need an assumption about the order of magnitude of the correlations at time zero. We should realize that we do not have much control on these functions and, when we consider correlations of, say, 32 particles, it becomes impossible to prepare the system, by external means, in a state where that function has a prescribed value. These correlations are actually created and destroyed by the internal mechanisms of interaction. Hence it is reasonable to assume that these functions have the same order in λ as the correlations which could be created from the vacuum by the interactions. The lowest order correlations are then binary correlations $\rho_{\underset{\sim}{k},-\underset{\sim}{k}}$, because these can be obtained from ρ_o by the action of $\lambda<\underset{\sim}{k},-\underset{\sim}{k}|L'_{ij}|0>\rho_o$: they are therefore of order λ; and the lowest order destruction term in eq. (3.10) is

$$\int_o^t d\tau VE(\tau)\, U^{o}(t-\tau)Cf(0) = \frac{\lambda}{2\pi} \sum_{ij} \int dz\ e^{-izt} <0|L'_{ij}|\underset{\sim}{k},-\underset{\sim}{k}> \frac{1}{i(\underset{\sim}{k}\underset{\sim}{v}_i-\underset{\sim}{k}\underset{\sim}{v}_j-z)} \rho_{\underset{\sim}{k},-\underset{\sim}{k}}(0) . \tag{4.3}$$

Let us now collect the partial results and write the master equation explicitly (using (1.21)(we have now taken $\lambda=1$)

$$\partial_t \rho_o(t) = \sum_{i<j} \int_o^t dt \frac{1}{2\pi} \int_c dz\ e^{-izt} \frac{8\pi^3}{\Omega} \sum_{k} \underset{\sim}{\partial}_{ij}\, kV_{\underset{\sim}{k}} \frac{1}{i(kv_i-kv_j-z)} V_k \underset{\sim}{k}\cdot\underset{\sim}{\partial}_{ij}\rho_o(t-\tau) -$$

$$+ \sum_{i<j} \frac{1}{2\pi} \int_c dz\ e^{-izt} \frac{8\pi^3}{\Omega} \sum_{\underset{\sim}{k}} \underset{\sim}{\partial}_{ij}\, \underset{\sim}{k}V_k \frac{1}{i(\underset{\sim}{k}\underset{\sim}{v}_i-\underset{\sim}{k}\underset{\sim}{v}_j-z)} \rho_{\underset{\sim}{k},-\underset{\sim}{k}}(0) . \tag{4.4}$$

We shall now examine the asymptotic behavior of this equation. Let us start with the destruction fragment which we write as

$$D(t) = \sum_{i<j} \underset{\sim}{\partial}_{ij} \frac{1}{2\pi i} \int dz\, e^{-izt} \underset{\sim}{\tilde{D}}_{ij}(z) \quad , \tag{4.5}$$

where

$$\underset{\sim}{\tilde{D}}_{ij}(z) = \frac{8\pi^3}{\Omega} \sum_{\underset{\sim}{k}} \frac{\underset{\sim}{k} V_k \rho_{\underset{\sim}{k},-\underset{\sim}{k}}(0)}{\underset{\sim}{k}(\underset{\sim}{v}_i - \underset{\sim}{v}_j) - z} \quad . \tag{4.6}$$

If we perform the Laplace transformation first, we get a sum of oscillating terms like $\exp[ik(\underset{\sim}{v}_i - \underset{\sim}{v}_j)t]$: D(t) is a periodic function of time and nothing here looks like our assumption (3.13).

Now, let us assume the system is very large. We know that the spectrum of allowed k-values then becomes continuous, and the sum over k becomes an integral, according to (1.22). But now, a dramatic change of properties occurs.

Indeed, whereas the original $\underset{\sim}{\tilde{D}}_{ij}(z)$ had a large number of poles on the real axis - each giving an oscillating contribution - the limiting value of $\underset{\sim}{\tilde{D}}_{ij}(z)$ no longer has any poles on the real axis. These poles have merged into each other and as a result the limiting function has a cut, i.e. a finite discontinuity along the real axis. It actually goes over into a so-called <u>Cauchy integral</u>:

$$\underset{\sim}{\tilde{D}}_{ij}(z) = \int d\underset{\sim}{k} \frac{\underset{\sim}{k} V_{\underset{\sim}{k}} \rho_{\underset{\sim}{k},-\underset{\sim}{k}}(0)}{\underset{\sim}{k}(\underset{\sim}{v}_i - \underset{\sim}{v}_j) - z} \quad . \tag{4.7}$$

These are well-known objects, and one can tell a lot about them without any calculations. In particular, $\underset{\sim}{\tilde{D}}_{ij}(z)$ is regular for z located anywhere in the upper half-plane. Although it has a discontinuity on the real axis, it can usually be continued analytically into the lower half-plane (This continuation looks of course different from (4.7)). Now, $\underset{\sim}{\tilde{D}}_{ij}(z)$ must have some singularities (otherwise it would be constant); as it is regular above the real axis, it can have singularities only in the lower half plane. Let us assume for simplicity that these singularities are simple poles. Their approximate location can be guessed from an intuitive argument. The distance from the real axis will indeed be determined by the form of the integrand in (4.7), i.e. by the properties of the potential V_k and of the initial correlations $\rho_{\underset{\sim}{k},-\underset{\sim}{k}}(0)$. But we already assumed the initial correlations to be created by the interactions, so the only characteristic factor left is V_k. This function can usually be characterized by a range of action r_o. If we also consider that most of the particles have velocities close to some average velocity v, we can construct a characteristic time $\tau_c = r_o/v$. Therefore, from a dimensional argument we can expect the poles of $\tilde{D}_{\underset{\sim}{ij}}(z)$ to lie at an effective distance τ_c^{-1} below the real axis. This argument is of course very rough, but at least in simple cases its conclusions can be checked explicitly on simple examples of V_k.

Let us now go back to eq. (4.5). If we now evaluate the Laplace transform, we see that D(t) is given by the residues at the poles of $\tilde{D}_{\underset{\sim}{ij}}(z)$; these will now give rise to exponentially decaying terms e^{-t/τ_c}. Hence, for times much longer than τ_c, $D(t)\to 0$, and we have justified the assumption (3.13). Let us summarize: the assumption (3.13) is valid under the following conditions:

a) large systems

b) interaction potentials of finite range

c) well behaved initial correlations; in particular correlations created by interactions.

Let us now pursue the analysis with the diagonal fragment in eq. (4.4), which we rewrite in the form

$$\Psi\rho_o = \sum_{i<j} \int_o^t dt \frac{1}{2\pi i} \int_C dz\, e^{-izt}\, \underset{\sim}{\partial}_{ij}\, \underset{\approx}{\Phi}_{ij}(z)\, \underset{\sim}{\partial}_{ij}\, \rho_o(t-\tau), \tag{4.8}$$

where, in the limit of a large system

$$\underset{\approx}{\Phi}_{ij}(z) = \int d\underset{\sim}{k} \frac{\underset{\sim}{k}\,\underset{\sim}{k}\, V_k^2}{\underset{\sim}{k}(\underset{\sim}{v}_i - \underset{\sim}{v}_j) - z} . \tag{4.9}$$

Let us first expand $\rho_o(t-\tau)$ around $\tau=0$

$$\Psi\rho_o = \sum_{i<j} \int_o^t d\tau \frac{1}{2\pi i} \int dz\, e^{-iz\tau} \underset{\sim}{\partial} \underset{\approx}{\Phi}(z)\partial\{\rho_o(t)-\tau \frac{\partial\rho_o}{\partial t} + \ldots\} . \tag{4.10}$$

Now, we note that $\frac{\partial\rho_o}{\partial t}$ is proportional to λ^2; on the other hand τ is cut by the memory kernel for values larger than τ_c. Hence the second term (and the following ones) in brackets are negligible in the present approximation. As a result, the equation has now become markoffian: the memory effects are negligible in a weakly coupled gas.

We can now perform the τ integration:

$$\Psi\rho_o = \sum_{i<j} \frac{1}{2\pi i} \int_C dz \frac{e^{-izt}-1}{-iz} \underset{\sim}{\partial} \underset{\approx}{\Phi}(z) \underset{\sim}{\partial} \rho_o(t) . \tag{4.11}$$

It is seen that the term -1 gives no contribution. The remaining Laplace transform is now evaluated:

$$\Psi\rho_o = \sum_{i<j} \underset{\sim}{\partial} \{ \Phi(0) + \sum_{\text{poles of } \underset{\approx}{\Phi}} e^{-iz_jt} \operatorname{Res} \frac{\Phi(z)}{-iz} \} \underset{\sim}{\partial} \rho_o(t) . \tag{4.12}$$

The arguments used in the discussion of the destruction fragment can be applied here too. Indeed, $\underset{\approx}{\Phi}(t)$ is also a Cauchy integral, and therefore its poles are way down in the lower half plane, at distances of order τ_c^{-1}. Therefore, if we consider times much longer than τ_c, the second term in brackets will be damped out and we are left only with the contribution of $\Phi(0)$. The latter can be written explicitly by noting that,

$$\lim_{z\to 0+} \frac{1}{x-z} = \pi i\delta_-(x) \equiv \pi i\delta(x) + P\frac{1}{x} .$$

We are then left with the final equation for the asymptotic distribution of velocities, $\rho_o(t)$:

$$\partial_t \rho_o(t) = \sum_{i<j} \int d\underset{\sim}{k}\, V_k^2\, \underset{\sim}{k}\underset{\sim}{\partial}_{ij}\, \pi\delta(\underset{\sim}{k}\underset{\sim}{v}_i-\underset{\sim}{k}\underset{\sim}{v}_j)\, \underset{\sim}{k}\underset{\sim}{\partial}_{ij}\, \rho_o(t) \equiv \Psi\rho_o . \tag{4.13}$$

This is the explicit form of the equation (3.14) for a homogeneous weakly coupled gas.

We note that this equation describes an irreversible approach to equilibrium. Indeed, the operator Ψ is <u>self-adjoint</u>, hence it has real, (and moreover, negative) eigenvalues. Therefore the solution of this equation is of the form

$$\rho_o(t) = \sum_n c_n e^{-\chi_n t} \psi_n \quad , \tag{4.14}$$

where $-\chi_n$ are the eigenvalues of Ψ and ψ_n the corresponding eigenfunctions. These eigenvalues introduce a new time-scale $\tau_{rn} = \chi_n^{-1}$; for small values of n, these <u>relaxation times</u> are much longer than τ_c (for a weakly coupled system). Moreover, the equation necessarily has an eigenvalue $\chi_o = 0$. The corresponding eigenfunction is the stationary solution of (4.13), i.e. the <u>equilibrium solution</u>. In this case it turns out to be

$$\rho_o^{eq} = \exp\,[-\rho_j \Sigma\, \rho_j^2/2m] \quad , \tag{4.15}$$

which is the Maxwell distribution of free particles: this is the endpoint of the evolution.

We conclude here the comments about the weakly coupled gas. The same type of calculations can be extended to the next orders in perturbation theory. If one does so, one immediately runs into difficulties. The origin of these difficulties can be seen in eq. (4.10). In higher orders, the time scales τ_c and τ_r are no longer widely separated; as a result one must retain some of the bracketted terms in eq. (4.10), to account for the finite duration of the collisions. After some rearrangement, the kinetic equation appears in the form

$$\partial_t \rho_o = \Omega \psi \rho_o \quad , \tag{4.16}$$

where ψ is a self-adjoint operator, but Ω is <u>not</u>.

Hence, in higher orders the irreversibility no longer appears as a pure phenomenon. In other words there is a mixture or a superposition of reversible and irreversible effects, and all this shows up in these very complicated equations. To understand these difficulties we could make the following supposition. Maybe the representation in terms of particles is not a convenient one. Maybe the behavior could become much simpler if, before writing down the equations, we could make some transformation to a new set of degrees of freedom. These would depend in a complicated way on the initial ones and could be called quasi-particles (using the terminology first introduced by Landau). If one could construct quasi-particles to describe "real" entities existing in the system, then one could hope that the kinetic equation, written in terms of quasi-particles, would again have such simple properties as the weak coupling equation. This was one of the guiding ideas in the latest work of our group in Brussels and I would like to devote my lecture tomorrow to a short discussion of these recent advances.

CHAPTER V. ASYMPTOTIC DYNAMICS

Let me warn you that in today's lecture, my purpose is to give you a feeling about the recent work in non-equilibrium statistical mechanics. I shall state a number of results, but in one hour it will be impossible to give any technical proofs.

Let us start again from equation (3.3). In the case of a homogeneous system, eqs. (1.15a) and (4.1) show us that $VR^o(z) \equiv (-iz)^{-1}$, and eq. (3.3) can then be written as

$$R(z) = \sum_{n=0}^{\infty}(V+CR^o(z)E(z)V)\frac{1}{(-iz)^{n+1}}[E(z)V]^n[V+VE(z)R^o(z)C]+CR^o+CR^oER^oC \ .$$

$$\text{(homogeneous)} \qquad (5.1)$$

Hence, each term in the series has a multiple pole in z=0. We have seen yesterday that this pole plays a quite important role in determining the asymptotic behavior of ρ_0. This same pole can actually be exhibited in the general case of inhomogeneous systems too, by a slight rearrangement of the series:

$$R(z) = \sum_{n=0}^{\infty}(V+CR^o(z)E(z)V)\frac{1}{(-iz)^{n+1}}[L^oV+E(z)V]^n[V+VE(z)R^o(z)C]$$

$$+ CR^o(z)+CR^o(z)E(z)R^o(z)C \ . \qquad (5.2)$$

We have seen that in weakly coupled systems, the only important contribution for long times comes from the residue in z=0. For strong coupling however, the two characteristic time scales are no longer well separated; therefore one can no longer claim that the contributions from the singularities of E(z) are small. We may however still <u>separate</u> the two types of contributions by writing

$$f(t) = \bar{f}(t) + \hat{f}(t), \qquad (5.3)$$

where $\bar{f}(t)$ is defined as the contribution from the residue in z=0 of the integrand in eq. (1.14):

$$\bar{f}(t) = \underset{z=0}{\mathrm{Res}}\{e^{-izt}R(z)\}f(0) \equiv \Sigma(t)f(0) . \qquad (5.4)$$

This residue can be easily calculated from (5.2):

$$\Sigma(t) = \sum_{n=0}^{\infty} \frac{1}{n!}\{(t+i\frac{\partial}{\partial t})^{n}(V+CR^{\circ}EV)(L^{\circ}V+EV)^{n}(V+VER^{\circ}C)\}_{z=0} . \qquad (5.5)$$

This expression can also be written in a very conveniently factorized form:

$$\Sigma(t) = (V+C\underline{\underline{C}}V)e^{\Omega\psi t}\ VAV[V+V\underline{D}C] , \qquad (5.6)$$

where $\underline{\underline{C}}$, Ω, ψ, A, $\underline{\underline{D}}$ are complicated but well defined operators, whose explicit expression we shall not write down.

We wish to show now that, although $\hat{f}(t) \equiv f(t)-\bar{f}(t)$ <u>cannot</u> be assumed to be <u>small</u> compared to $\bar{f}(t)$, it can be considered as <u>irrelevant</u> in many problems.

Let us therefore study some properties of $\bar{f}(t)$. Let us first write, in parallel to (5.3) a decomposition of the propagators

$$U(t) = \Sigma(t) + \hat{\Sigma}(t) , \qquad (5.7)$$

where U(t) is the exact Liouvillian propagator: exp(Lt). A first, fundamental property of $\Sigma(t)$ is the <u>semi-group property</u>:

$$\Sigma(t)\Sigma(t') = \Sigma(t+t'), \qquad t\geq 0,\ t'\geq 0 . \qquad (5.8)$$

Although it appears quite natural, this property is by no means trivial. An extremely interesting property is the fact that, as t=0, $\Sigma(0)$ does not reduce to the unit operator (as does U(0)), but rather to a well defined operator Π:

$$\Sigma(0) = \Pi . \qquad (5.9)$$

The definition of Π can be immediately read off from eq. (5.5) or (5.6). This fact is by no means inconsistent and is well-known in the theory of semi-groups. This operator Π has a very remarkable property which follows immediately from (5.8):

$$\Pi^2 = \Pi \quad . \tag{5.10}$$

In other words, Π is a <u>projection operator</u>.

A second remarkable property is expressed in the following property:

$$\Sigma(t) = \Pi U(t) = U(t)\Pi \ . \tag{5.11}$$

It then follows from (5.4) that $\bar{f}(t)$ is nothing other than the <u>projection</u> of the exact distribution function under the operator Π:

$$\bar{f}(t) = \Pi f(t) \quad , \tag{5.12}$$

and hence:

$$f(t) = \Pi f(t) + (1-\Pi)f(t) \ . \tag{5.13}$$

Let us now see the status of these two terms with respect to the classification into F and G classes, eqs. (2.3),(2.4). One can show:

$$\Pi f \in F \quad , \qquad (1-\Pi)f \in G \ . \tag{5.14}$$

Hence, the component Πf carries the whole normalization of the distribution.

Finally, a crucial theorem is the following:

$$\Pi f(H) = f(H) \quad , \tag{5.15}$$

where $f(H)$ is an arbitrary function of the hamiltonian. This means that, in particular, the equilibrium distribution lies entirely in the "Π-subspace." Hence, the component $\Pi f(t)$ evolves more or less smoothly towards equilibrium whereas $(1-\Pi)f(t)$ tends to zero as the system reaches equilibrium. As many important quantities in statistical mechanics are defined as functionals of H, this shows what we meant by saying that the component $\hat{f}(t)$ is in some sense "irrelevant" in many cases.

Let us now examine the relation between the two systems of projectors we have defined, viz. V and π.

A remarkable property can be immediately read off from eq. (5.6): the projections of $\bar{f}(t) \equiv \Sigma(t)f(0)$ on the vacuum and on the correlations are related:

$$C\Pi f(t) = \underline{\underline{C}}V\Pi f(t) . \tag{5.16}$$

Hence, in the "subspace" Π, there is no need for a separate study of the vacuum and of the correlations: the latter are functionals of the vacuum. This property is a very far-reaching generalization of Bogolyubov's ansatz on the functional dependence of the correlations on f_1. Moreover it represents a very remarkable "contraction" of the description of the systems, in the sense mentioned at the beginning of these lectures. Hence, the whole evolution in the Π-subspace is governed by the equation of evolution for the vacuum. The latter is easily obtained from (5.6)

$$\partial_t V\bar{f}(t) = \Omega\psi V\bar{f}(t). \tag{5.17}$$

This equation is the general kinetic equation, which we alluded to in yesterday's lecture. We should mention here that the appropriate initial condition for this equation is not the original $f(0)$, but rather:

$$V\bar{f}(0) \equiv V\Pi f(0) = VAV[V+VDC]f(0) . \tag{5.18}$$

It is easy to understand why we need a new initial condition for (5.17). The latter is an asymptotic equation, which gives a good approximation for $f(t)$ as $t\to\infty$. In order to obtain a solution of the approximate equation (5.17) which matches the exact solution for long times, we need to choose correctly the approximate trajectory right at the start (see Figure 2).

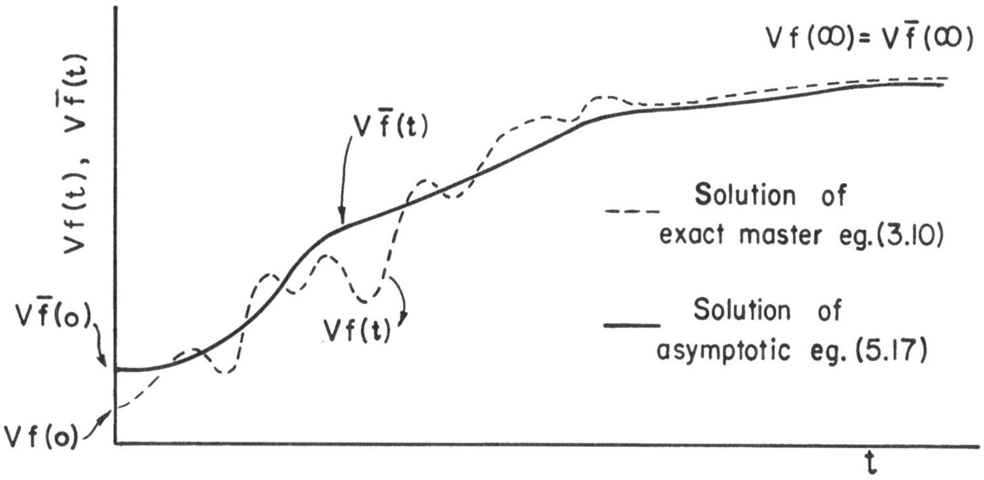

FIGURE 2

We have mentioned yesterday that eq. (5.17) leads to difficulties in interpretation. A very significant step further has been achieved when it was realized that there exists an operator χ which performs the following transformation:

$$V\Sigma(t)V = \chi\ e^{t\Phi}\ \chi^{+} \quad , \tag{5.19}$$

where χ^{+} is the adjoint of χ. Noting that, for homogeneous systems, $VDC = (C\underline{\underline{C}}V)^{+}$, it is then an easy matter to show that, on defining

$$f_R(t) = \chi^{-1} V\Pi f(t) \quad , \tag{5.20}$$

this "redefined" distribution function obeys the simpler equation

$$\partial_t f_R(t) = \Phi\, f_R(t) \quad . \tag{5.21}$$

One can see from (5.19) and (5.15) that

$$V\Sigma(0)V \equiv VAV = \chi\chi^{+} . \tag{5.22}$$

If χ is chosen to be a hermitian operator, then it would be given simply by the square root of the operator VAV, which is well-defined. One can then show that the operator Φ is a <u>hermitian operator</u>. Hence, our choice of χ realized the program of the quasi-particle transformation in the sense that $f_R(t)$ obeys a true kinetic equation collision operator with real eigenvalues.

Unfortunately this is not the end of the story. It is easily seen that if we multiply χ by an arbitrary unitary operator (say χ''), then all the relevant properties are still maintained. Hence the operator χ is not unique. To make the transformation unambiguous, we need an extra argument. This can be obtained by requiring that the equilibrium distribution of quasi-particles be given by a Maxwellian distribution, of the form:

$$f_R^{eq} = Z^{-1} \exp[-\beta H_R] \quad , \tag{5.23}$$

where

$$H_R = \chi^{-1} VH = \chi^{-1} H_o \quad ,$$

is a quasi-particle energy. This criterion, which is highly non-linear, is very strong. It excludes the hermitian solution we mentioned above (except in low orders of perturbation). An equation has been proposed for the construction of the correct χ. It has been tested that, up to order λ^8, for a variety of models, this equation gives the correct answer; however a general proof is still lacking and much effort is devoted at present to this problem.

With this glimpse into present-day research, I shall end these lectures. The short review given here is necessarily superficial. It is mainly intended to attract attention on this very active and exciting field of theoretical physics. People interested in further study of these problems are advised to read some of the references provided.

References

General references for non-equilibrium statistical mechanics according to the "Brussels school" are found in the following monographs and review papers.

[1] I. PRIGOGINE: Non-Equilibrium Statistical Mechanics (New York: Interscience-Wiley, 1962)

[2] R. BALESCU: Statistical Mechanics of Charged Particles (New York: Interscience-Wiley, 1963)

[3] P. RESIBOIS: in Physics of Many Particle Systems ed. by E. Meeron (New York: Gordon and Breach, 1966)

[4] I. PRIGOGINE: in Nato School on Non Linear Physics and Mathematics (Munich: Springer, 1968)

The vacuum and correlation projection operators have been introduced in two papers.

[1] R. BALESCU: Physica 38, 98 (1968)
Physica 42, 464 (1969)

The problems connected with the quasi-particle theory are presented in the following papers (among others).

[1] I. PRIGOGINE, F. HENIN AND C. GEORGE: Proc. Nat. Acad. Sci. USA 59, 7 (1968)

[2] C. GEORGE: Physica 37, 182 (1967)

[3] C. GEORGE: Bull. Classe Sci., Acad. Roy. Belg. 53, 623 (1967)

Lecture Notes in Physics

Bisher erschienen / Already published

Vol. 1: J. C. Erdmann, Wärmeleitung in Kristallen, theoretische Grundlagen und fortgeschrittene experimentelle Methoden. 1969. DM 20,–/ $ 5.50

Vol. 2: K. Hepp, Théorie de la renormalisation. 1969. DM 18,–/ $ 5.00

Vol. 3: A. Martin, Scattering Theory: Unitarity, Analyticity and Crossing. 1969. DM 14,–/ $ 3.90

Vol. 4: G. Ludwig, Deutung des Begriffs physikalische Theorie und axiomatische Grundlegung der Hilbertraumstruktur der Quantenmechanik durch Hauptsätze des Messens. 1970. DM 28,– / $ 7.70

Vol. 5: M. Schaaf, The Reduction of the Product of Two Irreducible Unitary Representations of the Proper Orthochronous Quantummechanical Poincaré Group. 1970. DM 14,– / $ 3.90

Vol. 6: Group Representations in Mathematics and Physics. Edited by V. Bargmann. 1970. DM 24,– / $ 6.60

Vol. 7: Lectures in Statistical Physics. 1971. DM 18,– / $ 5.00

Selected Issues from

Lecture Notes in Mathematics

Vol. 7: Ph. Tondeur, Introduction to Lie Groups and Transformation Groups. VIII, 176 pages. 1965. DM 13,50

Vol. 8: G. Fichera, Linear Elliptic Differential Systems and Eigenvalue Problems. IV, 176 pages. 1965. DM 13,50

Vol. 17: C. Müller, Spherical Harmonics. IV, 46 pages. 1966. DM 5,–

Vol. 25: R. Narasimhan, Introduction to the Theory of Analytic Spaces. IV, 143 pages. 1966. DM 10,–

Vol. 33: G. I. Targonski, Seminar on Functional Operators and Equations. IV, 110 pages. 1967. DM 10,–

Vol. 35: N. P. Bhatia and G. P. Szegö, Dynamical Systems. Stability Theory and Applications. VI, 416 pages. 1967. DM 24,–

Vol. 40: J. Tits, Tabellen zu den einfachen Lie Gruppen und ihren Darstellungen. VI, 53 Seiten. 1967. DM 6,80

Vol. 45: A. Wilansky, Topics in Functional Analysis. VI, 102 pages. 1967. DM 9,60

Vol. 52: D. J. Simms, Lie Groups and Quantum Mechanics. IV, 90 pages. 1968. DM 8,–

Vol. 55: D. Gromoll, W. Klingenberg und W. Meyer, Riemannsche Geometrie im Großen. VI, 287 Seiten. 1968. DM 20,–

Vol. 56: K. Floret und J. Wloka, Einführung in die Theorie der lokalkonvexen Räume. VIII, 194 Seiten. 1968. DM 16,–

Vol. 60: Seminar on Differential Equations and Dynamical Systems. Edited by G. S. Jones. VI, 106 pages. 1968. DM 9,60

Vol. 71: Séminaire Pierre Lelong (Analyse), Année 1967-1968. VI, 190 pages. 1968. DM 14,– / $ 3.90

Vol. 81: J.-P. Eckmann et M. Guenin, Méthodes Algébriques en Mécanique Statistique. VI, 131 pages. 1969. DM 12,–

Vol. 82: J. Wloka, Grundräume und verallgemeinerte Funktionen. VIII, 131 Seiten. 1969. DM 12,–

Vol. 91: N. N. Janenko, Die Zwischenschrittmethode zur Lösung mehrdimensionaler Probleme der mathematischen Physik. VIII, 194 Seiten. 1969. DM 16,80

Vol. 102: F. Stummel, Rand- und Eigenwertaufgaben in Sobolewschen Räumen. VIII, 386 Seiten. 1969. DM 20,–

Vol. 103: Lectures in Modern Analysis and Applications I. Edited by C. T. Taam. VII, 162 pages. 1969. DM 12,–

Vol. 104: G. H. Pimbley, Jr., Eigenfunction Branches of Nonlinear Operators and their Bifurcations. II, 128 pages. 1969. DM 10,–

Vol. 109: Conference on the Numerical Solution of Differential Equations. Edited by J. Ll. Morris. VI, 275 pages. 1969. DM 18,– / $ 5.00

Vol. 116: Séminaire Pierre Lelong (Analyse) Année 1969. IV, 195 pages. 1970. DM 14,– / $ 3.90

Vol. 126: P. Schapira, Théorie des Hyperfonctions. XI, 157 pages. 1970. DM 14,– / $ 3.90

Vol. 127: I. Stewart, Lie Algebras. IV, 97 pages. 1970. DM 10,– / $ 2.80

Vol. 128: M. Takesaki, Tomita's Theory of Modular Hilbert Algebras and its Applications. II, 123 pages. 1970. DM 10,– / $ 2,80

Vol. 140: Lectures in Modern Analysis and Applications II. Edited by C. T. Taam. VI, 119 pages. 1970. DM 10,– / $ 2.80

Vol. 144: Seminar on Differential Equations and Dynamical Systems, II. Edited by J. A. Yorke. VIII, 268 pages. 1970. DM 20,– / $ 5.50

Vol. 145: E. J. Dubuc, Kan Extensions in Enriched Category Theory. XVI, 173 pages. 1970. DM 16,– / $ 4.40

Vol. 155: J. Horváth, Several Complex Variables, Maryland 1970, I. V, 214 pages. 1970. DM 18,– / $ 5.00

Vol. 159: R. Ansorge und R. Hass, Konvergenz von Differenzverfahren für lineare und nicht lineare Anfangswertaufgaben. VIII, 145 Seiten. 1970. DM 14,– / $ 3.90

Beschaffenheit der Manuskripte

Die Manuskripte werden photomechanisch vervielfältigt; sie müssen daher in sauberer Schreibmaschinenschrift mit ausreichend großer Type geschrieben sein. Handschriftliche Formeln bitte nur mit schwarzer Tusche eintragen. Notwendige Korrekturen sind bei dem bereits geschriebenen Text entweder durch Überkleben des alten Textes vorzunehmen oder aber müssen die zu korrigierenden Stellen mit weißem Korrekturlack abgedeckt werden. Die reproduktionsfähigen Abbildungen (in Originalgröße) sollen in den Text eingeklebt werden. Falls das Manuskript oder Teile desselben neu geschrieben werden müssen, ist der Verlag bereit, dem Autor bei Erscheinen seines Bandes einen angemessenen Betrag zu zahlen. Die Autoren erhalten 50 Freiexemplare.

Zur Erreichung eines möglichst optimalen Reproduktionsergebnisses ist es erwünscht, daß bei der vorgesehenen Verkleinerung der Manuskripte der Text auf einer Seite in der Breite möglichst 18 cm und in der Höhe 26,5 cm nicht überschreitet. Entsprechende Satzspiegelvordrucke werden vom Verlag gern auf Anforderung zur Verfügung gestellt.

Manuskripte, in englischer, deutscher oder französischer Sprache abgefaßt, sind einzureichen bei: Springer-Verlag, 6900 Heidelberg, Postfach 1780.

Cette série a pour but de donner des informations rapides, de niveau élevé, sur des développements récents en physique, aussi bien dans la recherche que dans l'enseignement supérieur. On prévoit de publier.

1. des versions préliminaires de travaux originaux et de monographies
2. des cours spéciaux portant sur un domaine nouveau ou sur des aspects nouveaux de domaines classiques
3. des rapports de séminaires
4. des conférences faites lors de congrès ou de colloques

En outre il est prévu de publier dans cette série, si la demande le justifie, des rapports de séminaires et des cours multicopiés ailleurs mais déjà épuisés.

Dans l'intérêt d'une diffusion rapide, les contributions auront souvent un caractère provisoire; le cas échéant, les démonstrations ne seront données que dans les grandes lignes. Les travaux présentés pourront également paraître ailleurs. Une réserve suffisante d'exemplaires sera toujours disponible. En permettant aux personnes intéressées d'être informées plus rapidement, les éditeurs Springer espèrent, par cette série de «prépublications», rendre d'appréciables services aux instituts de physique. Les annonces dans les revues spécialisées, les inscriptions aux catalogues et les copyrights rendront plus facile aux bibliothèques la tâche de réunir une documentation complète.

Présentation des manuscrits

Les manuscrits, étant reproduits par procédé photomécanique, doivent être soigneusement dactylographiés type assez grand. Il est recommandé d'écrire à l'encre de Chine noire les formules non dactylographiées. Les corrections nécessaires doivent être effectuées soit par collage du nouveau texte sur l'ancien soit en recouvrant les endroits à corriger par du vernis correcteur blanc. Les illustrations; en dimension originale, préparées pour reproduction sont à insérer dans le texte. S'il s'avère nécessaire d'écrire de nouveau le manuscrit, soit complètement, soit en partie, la maison d'édition se déclare prête à verser à l'auteur, lors de la parution du volume, le montant des frais correspondants. Les auteurs recoivent 50 exemplaires gratuits.

Pour obtenir une reproduction optimale il est désirable que le texte dactylographié sur une page ne dépasse pas 26,5 cm en hauteur et 18 cm en largeur. Sur demande la maison d'edition met à la disposition des auteurs du papier spécialement préparé.

Les manuscrits en anglais, allemand ou français peuvent être adressés à Springer-Verlag, 6900 Heidelberg, Postfach 1780.